Series VS no.28, PP1 no. 24

Key population and vital statistics

Local and health authority areas

Population and vital statistics by area of usual residence in the United Kingdom, 2001

London: The Stationery Office

ISBN **0 11 621649 2**
ISSN **1469-2732**

Contact points

For enquiries about this publication, contact the Editor **Richard Pereira:**
Tel: 01329 813531
E-mail: subnatproj@ons.gov.uk

To order this publication, call The Stationery Office on **0870 600 5522.** See also back cover.

For general enquiries, contact the National Statistics Customer Enquiry Centre on 0845 601 3034
(minicom: 01633 812399)
e-mail: **info@statistics.gov.uk**
fax: 01633 652747
Letters: Room D115, Government Buildings, Cardiff Road, Newport, NP10 8XG

You can also find National Statistics on the internet - go to **www.statistics.gov.uk**

About the Office for National Statistics

The Office for National Statistics (ONS) is the government agency responsible for compiling, analysing and disseminating many of the United Kingdom's economic, social and demographic statistics, including the retail prices index, trade figures and labour market data, as well as the periodic census of the population and health statistics. The Director of ONS is also the National Statistician and the Registrar General for England and Wales, and the agency administers the statutory registration of births, marriages and deaths there.

A National Statistics publication

National Statistics are produced to high professional standards set out in the National Statistics Code of Practice. They undergo regular quality assurance reviews to ensure that they meet customer needs. They are produced free from any political interference.

Contents

Introduction

General

This publication draws together the United Kingdom's key population and vital statistics into a single volume. These statistics are produced by the Office for National Statistics (ONS), the General Register Office for Scotland (GROS), the Northern Ireland Statistics and Research Agency (NISRA) and the Scottish Health Service. The volume is published annually by the ONS. This is the fifth edition to include data for all the UK countries. Prior to this, only England and Wales had been included.

The statistics included in this volume cover:

- population;
- births and fertility rates;
- conceptions and conception rates;
- deaths and mortality rates;
- migration within the UK;
- migration between the UK and the rest of the world.

This introduction briefly describes the information presented in this publication. In order to help the reader interpret the data, some background to the calculations has also been included. More detailed explanations of the calculations used, the data sources and the definitions of terms are given in **Appendix A.**

Geographic presentation: local government and health authority areas

Data are presented for the constituent countries of the United Kingdom and at lower geographic levels, according to their local government organisation. An extensive restructuring of local government in Great Britain was carried out during the latter half of the 1990s.

In England, the traditional two-tier government structure in the non-metropolitan counties (district and county councils) was partially replaced by a phased introduction of single-tier unitary authorities. These changes started with the creation of the Isle of Wight Unitary Authority in April 1995 and was completed in April 1998 with the creation of 19 unitary authorities. Altogether, 46 unitary authorities were created in England over the four years, leaving 34 non-metropolitan (or shire) counties operating under a two-tier system.

The local government reorganisation did not involve the metropolitan areas: that is, the London Boroughs and metropolitan districts. These already operate as single-tier authorities, because they have no administrative upper-tier equivalent to county councils.

In Wales, the complete two-tier local government structure was replaced by 22 unitary authorities in April 1996. At the same time, the complete two tier structure in Scotland was replaced by 32 single-tier councils. Local government in Northern Ireland was last reorganised in 1973 when the two tier structure was replaced by single-tier district councils.

In addition to the changes to local government, the primary classification for the presentation of regional statistics in England was changed in 1996 from the Standard Statistical Regions to the Government Office Regions (GORs).

The changes are fully explained in the ONS publication, *Gazetteer of the old and new geographies of the United Kingdom* (ONS, London, 1999). A full version of this publication is available on the National Statistics website at **www.statistics.gov.uk/downloads/ons_geography/GazetteerRD_v3.pdf**. Details on the presentation of the areas are given in **Appendix B**. This appendix also gives an update of the changes in the reorganisation of health authorities and regional offices.

Data in this volume are presented for:

- the United Kingdom;
- England:
  - Government Office Regions;
  - unitary authorities;
  - metropolitan counties and districts;
  - non-metropolitan counties and districts;
  - Health Regional Office areas and health authorities;
- Wales:
  - unitary authorities;
  - health authorities;
- Scotland:
  - council areas;
  - health boards;
- Northern Ireland:
  - council districts (local government districts);
  - health and social services boards.

Population estimates

Sections 1-3 give estimates of the population at mid-2001; that is 30 June 2001. These are based upon 2001 Census results, and allow for subsequent births, deaths, migration, and ageing of the population.

The tables show the distribution of population in each area by sex and age-group at mid-2001. **Section 1** summarises the population, showing data for the constituent countries of the United Kingdom and Government Office Regions. **Section 2** is more detailed, giving population numbers for local and health authorities. **Section 3** shows how estimates of the population for larger areas have changed over the decade 1981-1991 and 1991-2001. Northern Ireland's population estimates from 1992-2000 have been revised in line with the 2001 Census figures. Changes are shown separately for natural change (that is, change due to births and deaths) and other change which includes change due to migration (together with other adjustments, where this is currently available).

Total estimated populations for the United Kingdom and its constituent countries at mid-2001 by single year of age and sex are shown for reference in **Annex A**.

Vital statistics

Section 4 provides the numbers of births and deaths in each area, and corresponding fertility and mortality rates, as well as related key statistics. The population estimates given in this section, and used in calculating rates, are the same as those presented in **Sections 1-3** and described above.

The tables in **Section 4** also include the number and proportion of births outside marriage in each area and the percentage of such births which are jointly registered by both parents resident at the same address. The latter figure gives an indication of the proportion of births outside marriage which are assumed to result from cohabiting relationships.

The tables also provide information on stillbirths and deaths in the first year of life and include a column showing the number of deaths at ages under one week. The *infant mortality rate* represents the number of deaths at ages under one year, per thousand live births. The *perinatal mortality rate* relates to the number of stillbirths plus the number of deaths at ages under seven days, per thousand live births and stillbirths. These rates, included in **Tables 4.1-4.3**, are widely used with other indicators as a measure of the standard of health in a community. They must, however, be used with care, particularly where they are based on a small number of deaths. Rates based on fewer than 20 cases are shown in italics. Wide fluctuations can occur from year to year in the perinatal and infant mortality rates for individual areas because of the small number of deaths involved. These mortality rates are based on all births without any adjustment for factors such as low birthweight and, therefore, should be interpreted with care.

The tables in **Section 4** also include a column showing low birthweight babies (weighing less than 2,500 grams) as a proportion of all live births for which a birthweight was stated[1]. The proportion of low birthweight babies can be used to help interpret perinatal and infant mortality rates, but it should be noted that the birthweight is not stated for a small proportion of live births. The figures in this volume are therefore an imperfect measure of the proportion of births of low birthweight in each area. Birthweight information on a registration basis is not available for Scotland, but can be obtained on a different basis from the Information and Statistics Division of the Scottish Health Service - see **Appendix D,** page 119, for full contact details. Birthweight information from Northern Ireland is not currently available, but may be included in future issues.

Appendix A contains the definitions of the Total Fertility Rate (TFR) and the Standardised Mortality Ratio (SMR). The SMR indicates how the mortality of a given area compares with the national level, after allowing for differences in age structure. For example, a SMR of 110 shows an area where mortality is 10 per cent above the national rate. The TFR indicates the average number of children that would be born to a woman if the current age-specific patterns of fertility persisted throughout her child-bearing life. Replacement level fertility is the level at which a population would be exactly replacing itself in the long-term, if mortality rates are constant and there is no net migration. It takes account of the sex ratio at birth (more boys are born than girls) and the age structure of the female population in the reproductive ages. In developed countries like the UK, a TFR of around 2.1 children is needed to ensure that the population replaces itself in the longer term.

Migration

Migration indicators are shown in **Section 5**.

As is usual after a Census, ONS has revised the past intercensal mid-year population estimates to be consistent with the 2001 Census. The results imply an average annual over-estimation of the net flow of migrants to the United Kingdom since 1991 of about 80 thousand. This figure was calculated by comparing the original population estimates (based on the 1991 Census) with final rebased population estimates taking account of 2001 Census results. The Migration Statistics Unit in ONS is therefore in the process of revising all migration estimates for 1992-2001 and full results will be published in late Summer.

---

[1] These figures are comparable with those published in this volume from 1989. Prior to 1989, the figures related to both live births and stillbirths, and births where the birthweight was unknown were included in the denominator.

The migration tables published in this volume are therefore subject to revision and have been published to provide users with a provisional migration data series to be used until such time as the revised figures are available. They are not consistent with the estimated components of population change shown in Section 3. The revised migration figures will be made available on the ONS website (www.statistics.gov.uk) in late Summer. Printed copies will also be available on demand and will be free of charge (email: migstatsunit@ons.gov.uk or tel: 01329 813897).

The figures in the tables at Section 5 are derived from the National Health Service Central Register (NHSCR) and health authority patient registers for migration within the United Kingdom.

A revised methodology for producing Total International Migration estimates is still being developed. No estimates for Total International Migration have therefore been included in this volume, but will be included in revised tables on the web and in print on demand copies.

Any migrating move has an origin and a destination. NHSCR data are recorded at the health authority level in England and Wales, or health board level in Scotland and Northern Ireland. Consequently, migration moves can be identified only if the move crosses the boundary between one health authority and another. Health authorities are therefore the lowest geographical level at which NHSCR-based migration estimates are available.

However, the patient registers held by the health authorities themselves contain postcode information, so additional estimates of migration within the UK are now available using this source at both health and local authority levels.

Migration data can be aggregated from local or health authorities to larger areas. In this volume the migration data are presented at the following levels:

- The United Kingdom;
- constituent countries of the United Kingdom;
- Government Office Regions;
- metropolitan counties and the remainder of Government Office Regions;
- health authorities;
- local authorities.

Table 5.1a shows numbers migrating to or from health authorities and health regional office areas. **Table 5.1b** presents migration in and out of local authorities, unitary authorities (UAs), counties and Government Office Regions. Both parts of **Table 5.1** are split by sex and broad age-group. A matrix of origins and destinations of moves is given in **Table 5.2a** and **5.2b** for the larger aggregated areas. **Table 5.3** shows migration from or to the countries of the UK and regions of England, to or from elsewhere in the UK.

It is important to note that migration for the larger aggregated areas (UK, countries and GORs) is presented in two ways:

- including migration within areas, (which, of course, excludes migration within the constituent local or health authorities);
- migration between (the larger aggregated) areas only.

This complication means that some care is needed when comparing figures. For example, the figures in **Table 5.1a** for both inflow and outflow for the constituent health authorities of the Northern and Yorkshire Regional Office sum to more than the flows for the regional office itself. This is because the figures for the regional office exclude internal moves between its constituent health authorities. In the net column however, these internal moves cancel each other out and so the net total of the constituent health authorities equals the net figure shown for the regional office. **Tables 5.2a** and **5.2b** show totals for areas with these internal moves both included and excluded.

Availability of data

The data contained in this volume can be obtained in electronic form from the appropriate contact points detailed in **Appendix D**. In some cases, more detailed data are available, and can also be requested. **Appendix C** lists publications of interest. Further information can also be found on the National Statistics web-site on **www.statistics.gov.uk**

Note on rounding

Totals presented in the tables may not necessarily agree with the sum of their components due to rounding.

Map 1 Counties and unitary authorities in England, Wales and Scotland, and Northern Irish district council areas 1 April 2001

Map 2a Proportion of the population at mid-2001 in certain age groups: counties/unitary authorities of England, Wales and Scotland, and Northern Irish district council areas: children 0-15

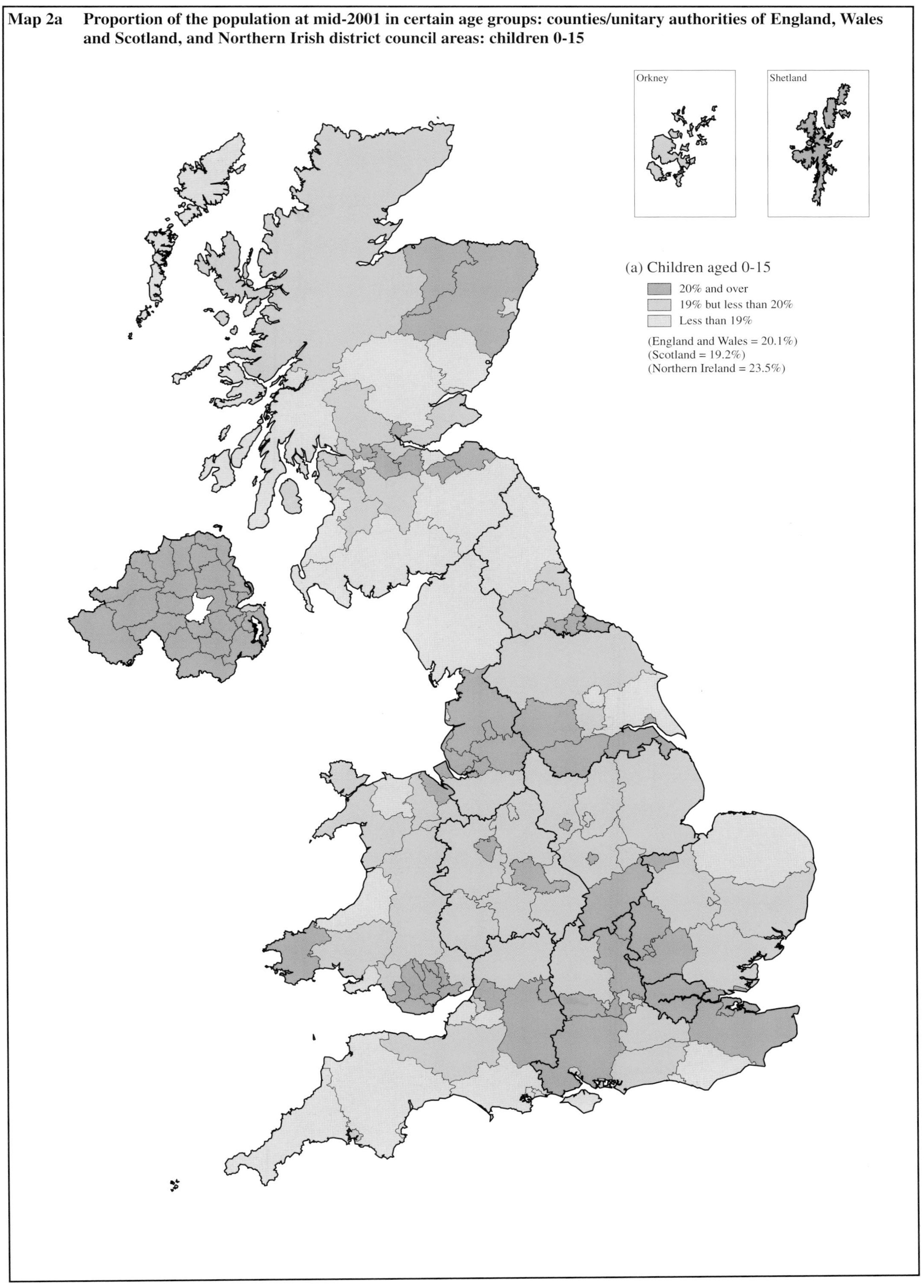

Map 2b Proportion of the population at mid-2001 in certain age groups: counties/unitary authorities of England, Wales and Scotland, and Northern Irish district council areas: working ages

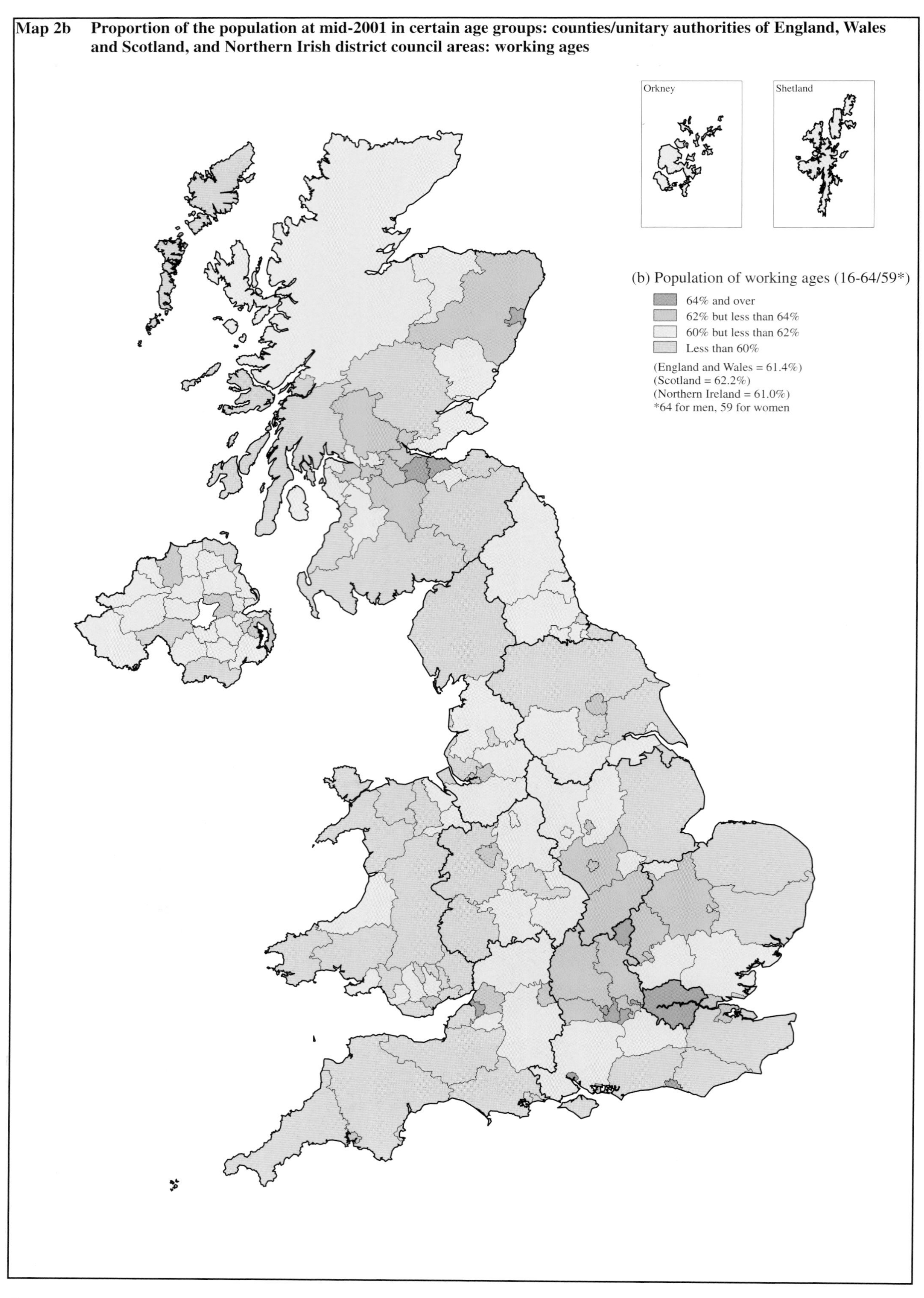

Map 2c Proportion of the population at mid-2001 in certain age groups: counties/unitary authorities of England, Wales and Scotland, and Northern Irish district council areas: retirement ages

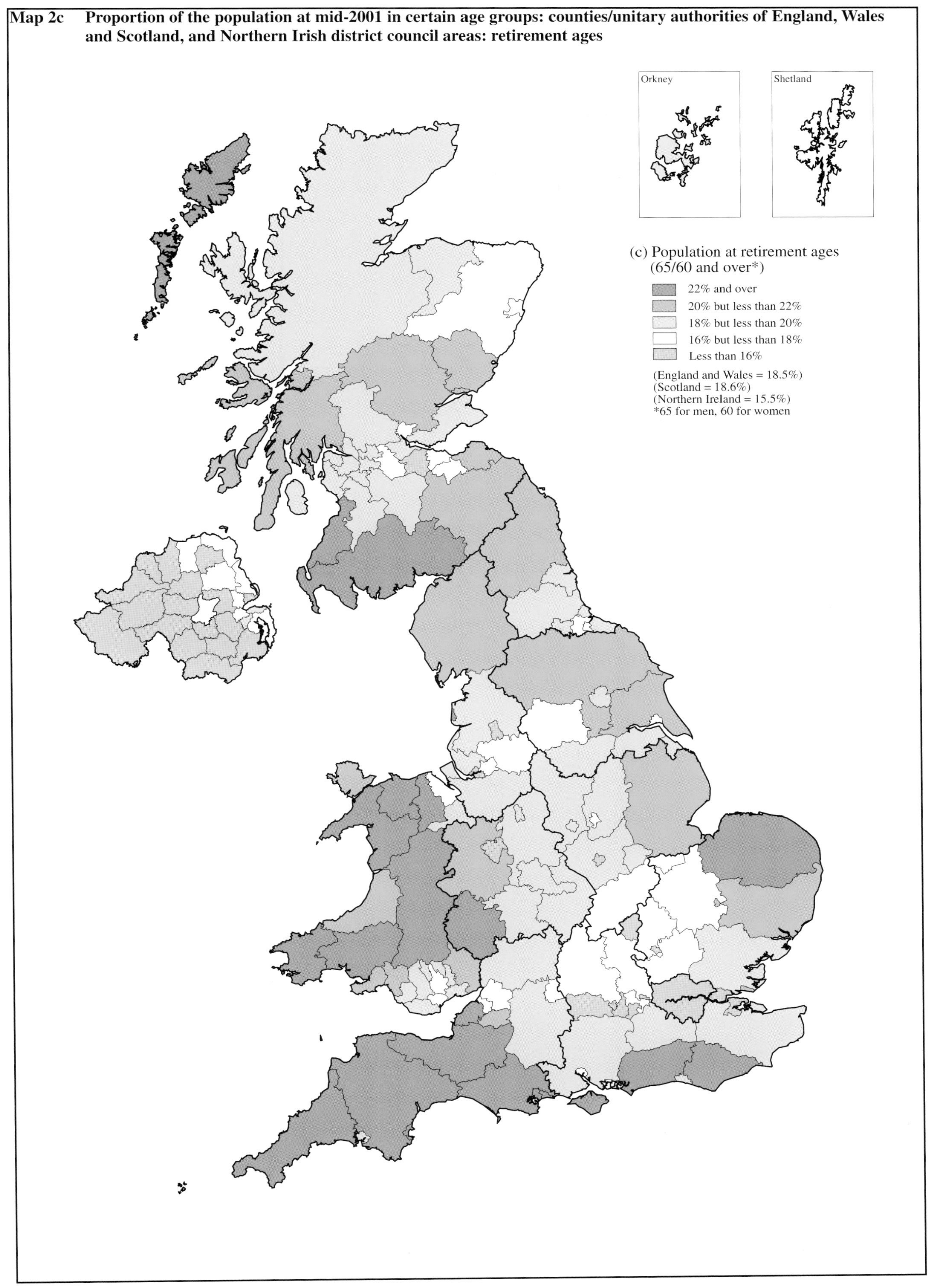

Table 1 Mid-2001 population estimates: estimated resident population by broad age-group and sex

Area	All ages		Under 1		1 - 4		5 - 15		16 - 29	
	Males	Females	Males	Females	Males	Females	Males	Females	Males	Females
United Kingdom	**28,611.3**	**30,225.3**	**338.4**	**324.0**	**1,442.6**	**1,372.3**	**4,292.4**	**4,085.0**	**5,142.5**	**5,166.7**
England and Wales	**25,354.7**	**26,729.8**	**300.9**	**287.9**	**1,279.2**	**1,218.2**	**3,792.1**	**3,608.8**	**4,540.1**	**4,559.3**
England	**23,950.6**	**25,230.8**	**284.6**	**272.6**	**1,209.8**	**1,152.1**	**3,576.7**	**3,404.6**	**4,301.4**	**4,317.0**
Government Office Regions										
North East	1,219.3	1,297.2	13.1	12.6	57.6	54.7	183.6	174.8	212.4	215.8
North West	3,260.5	3,471.0	37.9	36.4	163.9	155.5	510.3	486.3	567.0	581.2
Yorkshire and the Humber	2,413.8	2,553.4	28.1	26.8	120.4	115.8	369.5	353.8	431.9	435.2
East Midlands	2,050.4	2,124.7	22.9	21.7	99.6	94.1	307.9	290.7	355.8	348.1
West Midlands	2,575.5	2,691.6	30.4	29.4	132.8	125.4	397.5	379.6	453.3	450.5
East	2,642.2	2,752.7	31.1	30.0	133.5	126.6	389.9	372.6	447.0	439.8
London	3,479.5	3,708.5	49.6	47.4	194.1	186.9	496.7	475.8	757.5	800.8
South East	3,909.6	4,097.3	45.8	43.7	196.0	186.1	577.4	545.4	678.9	664.8
South West	2,399.8	2,534.4	25.8	24.5	111.9	106.9	343.9	325.6	397.6	380.8
Wales	**1,404.1**	**1,499.1**	**16.3**	**15.3**	**69.4**	**66.1**	**215.4**	**204.2**	**238.7**	**242.3**
Scotland	**2,433.7**	**2,630.5**	**26.3**	**25.7**	**115.4**	**108.8**	**355.5**	**338.6**	**438.5**	**444.0**
Northern Ireland	**824.4**	**864.9**	**11.1**	**10.4**	**47.9**	**45.3**	**144.8**	**137.6**	**164.6**	**163.4**

United Kingdom, by countries and, within England, Government Office Regions
thousands

30-44		45-59		60-64		65-74		75 and over		Area
Males	Females	Males	Females	Males	Females	Males	Females	Males	Females	
6,544.1	**6,747.2**	**5,520.0**	**5,624.2**	**1,409.3**	**1,470.4**	**2,303.8**	**2,636.0**	**1,618.2**	**2,799.6**	**United Kingdom**
5,797.0	**5,955.7**	**4,892.6**	**4,981.8**	**1,249.2**	**1,295.3**	**2,047.7**	**2,322.1**	**1,455.8**	**2,500.9**	**England and Wales**
5,502.1	**5,644.6**	**4,610.6**	**4,693.1**	**1,173.9**	**1,217.3**	**1,924.0**	**2,181.6**	**1,367.4**	**2,347.8**	**England**
										Government Office Regions
271.7	284.8	242.9	244.5	63.6	67.4	106.3	123.2	68.1	119.5	North East
728.9	759.0	638.5	646.7	167.6	174.8	268.5	310.1	177.9	321.0	North West
540.2	555.3	469.7	473.0	121.0	126.6	196.2	227.9	136.9	239.0	Yorkshire and the Humber
461.9	471.6	410.2	410.4	102.9	104.8	169.1	187.3	120.0	195.8	East Midlands
570.3	582.0	502.7	504.4	132.1	134.7	211.8	238.2	144.5	247.4	West Midlands
603.0	607.4	524.7	533.7	131.6	135.0	221.4	244.6	159.9	262.9	East
909.5	939.7	561.1	594.9	136.8	145.4	219.0	249.1	155.1	268.7	London
897.6	912.9	775.4	786.1	190.0	196.2	313.4	355.5	235.1	406.6	South East
518.9	532.0	485.3	499.5	128.2	132.5	218.4	245.6	169.9	286.9	South West
294.9	**311.1**	**282.1**	**288.6**	**75.3**	**77.9**	**123.7**	**140.5**	**88.3**	**153.0**	**Wales**
563.4	**599.9**	**483.1**	**496.1**	**124.6**	**136.9**	**200.5**	**246.1**	**126.2**	**234.3**	**Scotland**
184.5	**191.5**	**144.1**	**146.4**	**35.5**	**38.2**	**55.6**	**67.8**	**36.2**	**64.4**	**Northern Ireland**

Table 2.1 Mid-2001 population estimates: estimated resident population by broad age-group and sex

Area	All ages		Under 1		1 - 4		5 - 15		16 - 29	
	Males	Females	Males	Females	Males	Females	Males	Females	Males	Females
UNITED KINGDOM	**28,611.3**	**30,225.3**	**338.4**	**324.0**	**1,442.6**	**1,372.3**	**4,292.4**	**4,085.0**	**5,142.5**	**5,166.7**
ENGLAND AND WALES	**25,354.7**	**26,729.8**	**300.9**	**287.9**	**1,279.2**	**1,218.2**	**3,792.1**	**3,608.8**	**4,540.1**	**4,559.3**
ENGLAND	**23,950.6**	**25,230.8**	**284.6**	**272.6**	**1,209.8**	**1,152.1**	**3,576.7**	**3,404.6**	**4,301.4**	**4,317.0**
NORTH EAST	**1,219.3**	**1,297.2**	**13.1**	**12.6**	**57.6**	**54.7**	**183.6**	**174.8**	**212.4**	**215.8**
Darlington UA	**47.0**	**50.9**	**0.5**	**0.5**	**2.3**	**2.4**	**7.2**	**6.9**	**7.4**	**7.8**
Hartlepool UA	**42.6**	**46.1**	**0.5**	**0.5**	**2.2**	**2.1**	**7.1**	**6.9**	**6.8**	**7.3**
Middlesbrough UA	**64.7**	**70.1**	**0.8**	**0.8**	**3.4**	**3.3**	**10.9**	**10.7**	**12.6**	**13.0**
Redcar and Cleveland UA	**67.1**	**72.0**	**0.7**	**0.7**	**3.2**	**3.0**	**10.7**	**10.3**	**10.5**	**11.0**
Stockton-on-Tees UA	**87.2**	**91.4**	**1.0**	**0.9**	**4.4**	**4.0**	**14.1**	**13.2**	**14.9**	**15.1**
Durham	**240.0**	**253.7**	**2.4**	**2.3**	**10.9**	**10.5**	**34.9**	**33.3**	**40.9**	**41.1**
Chester-le-Street	26.1	27.6	0.3	0.3	1.2	1.2	3.7	3.6	4.0	4.0
Derwentside	41.3	43.9	0.4	0.4	1.9	1.9	6.0	5.7	6.5	6.6
Durham	43.2	44.6	0.4	0.4	1.7	1.6	5.3	5.0	10.3	10.2
Easington	45.5	48.5	0.5	0.5	2.2	2.2	7.3	6.8	7.3	7.6
Sedgefield	42.3	44.9	0.5	0.4	2.0	1.9	6.5	6.2	6.6	6.8
Teesdale	12.1	12.3	0.1	0.1	0.5	0.4	1.6	1.5	2.0	1.5
Wear Valley	29.5	31.9	0.3	0.3	1.4	1.3	4.4	4.3	4.3	4.5
Northumberland	**150.0**	**157.4**	**1.4**	**1.4**	**6.6**	**6.3**	**21.7**	**20.4**	**22.7**	**21.8**
Alnwick	15.1	16.0	0.1	0.2	0.7	0.6	2.0	2.0	2.0	2.0
Berwick-upon-Tweed	12.5	13.5	0.1	0.1	0.5	0.5	1.7	1.6	1.7	1.6
Blyth Valley	39.6	41.7	0.5	0.4	1.9	1.8	6.0	5.7	6.7	6.9
Castle Morpeth	24.4	24.6	0.2	0.2	1.0	0.9	3.3	3.1	3.6	3.0
Tynedale	28.7	30.2	0.3	0.3	1.2	1.1	4.2	4.1	4.0	3.6
Wansbeck	29.8	31.3	0.3	0.3	1.3	1.3	4.4	4.0	4.7	4.8
Tyne and Wear (Met County)	**520.7**	**555.7**	**5.8**	**5.5**	**24.6**	**23.2**	**77.1**	**73.2**	**96.6**	**98.6**
Gateshead	92.5	98.7	1.1	1.0	4.4	4.2	13.4	12.8	15.3	15.6
Newcastle upon Tyne	125.6	134.0	1.5	1.4	5.8	5.5	17.8	16.6	28.7	29.7
North Tyneside	91.9	100.1	1.0	1.0	4.3	4.1	13.4	13.0	14.5	15.3
South Tyneside	74.1	78.7	0.7	0.7	3.6	3.3	11.6	10.9	12.2	12.2
Sunderland	136.6	144.2	1.5	1.3	6.5	6.0	20.9	19.8	25.9	25.8
NORTH WEST	**3,260.5**	**3,471.0**	**37.9**	**36.4**	**163.9**	**155.5**	**510.3**	**486.3**	**567.0**	**581.2**
Blackburn with Darwen UA	**67.4**	**70.2**	**1.0**	**0.9**	**4.4**	**4.2**	**12.5**	**11.6**	**12.5**	**12.8**
Blackpool UA	**68.7**	**73.5**	**0.7**	**0.7**	**3.3**	**3.1**	**9.8**	**9.3**	**10.3**	**10.8**
Halton UA	**57.1**	**61.1**	**0.7**	**0.7**	**2.9**	**2.9**	**9.4**	**9.0**	**10.3**	**10.8**
Warrington UA	**94.0**	**97.2**	**1.1**	**1.0**	**4.7**	**4.8**	**14.4**	**13.7**	**15.6**	**15.2**
Cheshire	**328.3**	**346.0**	**3.6**	**3.5**	**15.8**	**15.0**	**48.9**	**46.0**	**50.2**	**51.0**
Chester	57.0	61.2	0.7	0.6	2.6	2.5	7.9	7.5	9.6	10.5
Congleton	44.5	46.2	0.4	0.5	2.1	2.0	6.4	6.0	7.0	6.7
Crewe and Nantwich	54.5	56.7	0.6	0.6	2.7	2.7	8.4	7.9	8.5	8.8
Ellesmere Port and Neston	39.8	41.8	0.4	0.4	2.1	1.9	6.5	5.8	6.2	6.1
Macclesfield	72.5	77.6	0.8	0.7	3.4	3.2	10.4	9.9	9.9	9.9
Vale Royal	60.0	62.3	0.7	0.7	3.0	2.8	9.3	8.8	9.1	8.9
Cumbria	**238.0**	**249.8**	**2.4**	**2.3**	**10.6**	**9.9**	**34.4**	**32.2**	**35.7**	**35.4**
Allerdale	45.6	48.0	0.4	0.4	2.0	1.9	6.6	6.2	6.8	6.5
Barrow-in-Furness	35.1	36.9	0.4	0.4	1.8	1.6	5.5	5.3	5.3	5.6
Carlisle	48.7	52.0	0.5	0.5	2.1	2.1	7.1	6.6	8.0	8.2
Copeland	34.5	34.7	0.3	0.3	1.5	1.5	5.2	4.7	5.4	5.2
Eden	24.6	25.3	0.3	0.2	1.1	1.0	3.3	3.2	3.4	3.3
South Lakeland	49.5	52.9	0.4	0.4	2.1	1.8	6.7	6.1	6.8	6.7

United Kingdom by countries and, within England, Government Office Regions; counties/unitary authorities, districts and London boroughs in Great Britain/district council areas in Northern Ireland

thousands

30-44		45-59		60-64		65-74		75 and over		Area
Males	Females	Males	Females	Males	Females	Males	Females	Males	Females	
6,544.1	**6,747.2**	**5,520.0**	**5,624.2**	**1,409.3**	**1,470.4**	**2,303.8**	**2,636.0**	**1,618.2**	**2,799.6**	**UNITED KINGDOM**
5,797.0	**5,955.7**	**4,892.6**	**4,981.8**	**1,249.2**	**1,295.3**	**2,047.7**	**2,322.1**	**1,455.8**	**2,500.9**	**ENGLAND AND WALES**
5,502.1	**5,644.6**	**4,610.6**	**4,693.1**	**1,173.9**	**1,217.3**	**1,924.0**	**2,181.6**	**1,367.4**	**2,347.8**	**ENGLAND**
271.7	**284.8**	**242.9**	**244.5**	**63.6**	**67.4**	**106.3**	**123.2**	**68.1**	**119.5**	**NORTH EAST**
10.7	**11.3**	**9.6**	**9.6**	**2.5**	**2.6**	**4.0**	**4.7**	**2.8**	**5.2**	**Darlington UA**
9.4	**10.3**	**8.3**	**8.3**	**2.2**	**2.4**	**3.9**	**4.4**	**2.2**	**3.9**	**Hartlepool UA**
13.8	**15.3**	**11.7**	**11.9**	**3.0**	**3.3**	**5.2**	**6.1**	**3.2**	**5.6**	**Middlesbrough UA**
14.3	**15.4**	**14.0**	**14.1**	**4.0**	**4.0**	**6.0**	**6.8**	**3.8**	**6.7**	**Redcar and Cleveland UA**
20.1	**21.3**	**17.2**	**17.2**	**4.3**	**4.4**	**7.1**	**8.1**	**4.2**	**7.2**	**Stockton-on-Tees UA**
54.1	**55.6**	**49.4**	**49.9**	**13.2**	**13.6**	**20.9**	**24.1**	**13.3**	**23.4**	**Durham**
6.4	6.6	5.6	5.6	1.5	1.5	2.2	2.6	1.3	2.2	Chester-le-Street
9.5	9.7	8.6	8.6	2.3	2.4	3.7	4.2	2.4	4.5	Derwentside
9.4	9.2	8.6	8.7	2.2	2.3	3.3	3.8	2.1	3.6	Durham
10.3	10.7	8.7	8.9	2.4	2.6	4.3	4.9	2.5	4.3	Easington
9.4	10.0	8.9	9.1	2.3	2.4	3.7	4.3	2.4	4.0	Sedgefield
2.5	2.5	2.7	2.7	0.7	0.8	1.2	1.3	0.8	1.4	Teesdale
6.6	7.0	6.4	6.3	1.7	1.7	2.6	3.1	1.7	3.4	Wear Valley
32.7	**33.9**	**33.4**	**33.7**	**8.4**	**8.9**	**13.9**	**15.6**	**9.4**	**15.3**	**Northumberland**
3.3	3.4	3.4	3.4	0.9	1.0	1.6	1.7	1.1	1.7	Alnwick
2.5	2.7	2.8	2.9	0.8	0.9	1.4	1.6	1.0	1.7	Berwick-upon-Tweed
8.8	9.2	8.7	8.9	2.0	2.1	3.0	3.4	2.0	3.3	Blyth Valley
5.1	5.1	5.5	5.5	1.5	1.5	2.5	2.7	1.7	2.6	Castle Morpeth
6.2	6.5	6.7	6.7	1.6	1.7	2.6	3.0	1.9	3.2	Tynedale
6.8	7.0	6.2	6.3	1.6	1.7	2.6	3.1	1.8	2.9	Wansbeck
116.6	**121.8**	**99.3**	**99.8**	**26.1**	**28.2**	**45.3**	**53.5**	**29.3**	**52.0**	**Tyne and Wear (Met County)**
21.2	22.2	18.0	18.2	5.1	5.4	8.7	9.9	5.2	9.4	Gateshead
27.5	28.6	22.0	21.5	5.4	6.2	9.8	11.8	7.1	12.7	Newcastle upon Tyne
21.0	22.2	18.9	19.1	4.7	5.2	8.4	10.0	5.7	10.3	North Tyneside
16.4	17.4	14.2	14.1	3.9	4.0	6.7	8.1	4.6	7.9	South Tyneside
30.4	31.4	26.1	26.9	6.9	7.4	11.7	13.7	6.7	11.8	Sunderland
728.9	**759.0**	**638.5**	**646.7**	**167.6**	**174.8**	**268.5**	**310.1**	**177.9**	**321.0**	**NORTH WEST**
14.7	**15.1**	**11.9**	**11.7**	**2.9**	**3.0**	**4.5**	**5.3**	**3.0**	**5.5**	**Blackburn with Darwen UA**
15.1	**15.1**	**13.9**	**13.9**	**4.2**	**4.3**	**6.6**	**7.6**	**4.8**	**8.9**	**Blackpool UA**
12.6	**13.6**	**11.7**	**12.1**	**2.7**	**2.7**	**4.3**	**4.8**	**2.5**	**4.4**	**Halton UA**
23.0	**23.2**	**18.7**	**18.8**	**4.8**	**5.0**	**7.1**	**7.7**	**4.5**	**7.8**	**Warrington UA**
74.5	**77.0**	**69.7**	**70.0**	**17.9**	**18.6**	**28.4**	**32.1**	**19.3**	**32.8**	**Cheshire**
12.7	13.1	11.8	11.9	3.1	3.3	5.0	5.8	3.6	6.0	Chester
10.1	10.3	9.9	10.0	2.5	2.5	3.7	4.0	2.5	4.2	Congleton
12.5	12.6	11.3	11.0	2.9	2.9	4.5	5.1	3.1	5.2	Crewe and Nantwich
9.0	9.5	7.8	8.1	2.2	2.4	3.5	4.1	2.1	3.6	Ellesmere Port and Neston
16.4	17.3	16.0	16.4	4.1	4.3	6.8	7.6	4.7	8.2	Macclesfield
13.8	14.3	12.9	12.6	3.1	3.2	4.8	5.5	3.3	5.5	Vale Royal
52.7	**53.0**	**50.6**	**50.8**	**14.0**	**14.2**	**22.6**	**25.3**	**15.1**	**26.6**	**Cumbria**
9.9	10.2	9.9	10.0	2.7	2.8	4.4	4.9	2.9	5.0	Allerdale
7.9	7.9	7.2	7.1	2.0	2.0	3.0	3.3	2.0	3.8	Barrow-in-Furness
11.0	11.1	9.9	10.1	2.6	2.8	4.5	5.2	3.0	5.3	Carlisle
8.0	7.8	7.2	6.8	1.9	2.0	3.1	3.4	1.8	3.1	Copeland
5.5	5.4	5.5	5.5	1.5	1.5	2.5	2.6	1.6	2.7	Eden
10.3	10.7	11.0	11.3	3.2	3.2	5.2	5.9	3.8	6.7	South Lakeland

Table 2.1 - *continued*

Area	All ages		Under 1		1 - 4		5 - 15		16 - 29	
	Males	Females	Males	Females	Males	Females	Males	Females	Males	Females
Greater Manchester (Met County)	**1,208.9**	**1,273.9**	**14.6**	**14.1**	**63.2**	**60.3**	**191.6**	**182.5**	**224.9**	**228.2**
Bolton	127.3	134.1	1.6	1.5	7.0	6.7	20.2	19.8	22.9	22.5
Bury	88.0	92.7	1.0	1.0	4.5	4.5	14.3	13.5	14.5	14.5
Manchester	191.8	201.1	2.5	2.5	10.0	9.6	29.7	28.3	50.9	52.0
Oldham	105.2	112.3	1.5	1.4	6.4	6.0	18.0	16.8	17.8	19.0
Rochdale	99.7	105.5	1.3	1.2	5.7	5.5	17.0	16.1	17.2	17.9
Salford	106.1	109.8	1.2	1.2	5.3	4.7	16.2	15.3	20.8	20.5
Stockport	137.3	147.3	1.5	1.5	6.8	6.6	21.1	20.2	21.0	21.9
Tameside	103.4	109.7	1.2	1.2	5.3	5.1	16.7	16.0	17.1	17.9
Trafford	102.2	108.0	1.1	1.1	5.0	4.7	15.7	14.8	17.1	16.9
Wigan	147.9	153.6	1.7	1.7	7.3	6.8	22.6	21.6	25.6	25.1
Lancashire	**551.0**	**584.8**	**6.5**	**6.0**	**27.2**	**25.5**	**85.0**	**81.7**	**93.2**	**95.4**
Burnley	43.4	46.1	0.5	0.5	2.4	2.2	7.6	7.1	7.2	7.8
Chorley	50.0	50.5	0.6	0.5	2.3	2.2	7.3	7.1	8.4	7.7
Fylde	35.2	38.2	0.4	0.3	1.5	1.4	4.7	4.7	4.9	4.4
Hyndburn	39.8	41.7	0.5	0.5	2.2	2.1	6.8	6.4	6.6	6.9
Lancaster	64.2	69.9	0.8	0.7	3.0	2.8	9.1	8.6	13.2	14.1
Pendle	43.5	45.7	0.6	0.6	2.5	2.3	7.3	7.0	7.5	7.6
Preston	63.1	66.5	0.7	0.8	3.3	3.0	9.7	9.5	13.0	14.2
Ribble Valley	26.4	27.6	0.3	0.3	1.2	1.2	4.0	3.8	3.7	3.6
Rossendale	32.0	33.7	0.4	0.4	1.6	1.7	5.4	5.2	5.1	5.2
South Ribble	50.7	53.3	0.6	0.5	2.4	2.2	7.8	7.5	8.2	8.0
West Lancashire	52.3	56.2	0.6	0.5	2.7	2.4	8.1	7.8	8.4	8.9
Wyre	50.3	55.4	0.6	0.5	2.2	2.0	7.3	7.2	7.1	7.0
Merseyside (Met County)	**647.2**	**714.4**	**7.4**	**7.1**	**31.9**	**30.0**	**104.3**	**100.3**	**114.4**	**121.6**
Knowsley	71.1	79.3	0.9	0.9	4.0	3.6	12.6	12.4	12.3	13.1
Liverpool	209.8	229.4	2.5	2.3	10.2	9.7	32.7	30.9	44.5	49.3
St Helens	85.7	91.1	1.0	0.9	4.3	3.9	13.5	13.1	14.3	14.4
Sefton	133.4	149.5	1.3	1.4	6.2	5.9	21.5	20.8	20.6	20.7
Wirral	147.1	165.1	1.7	1.7	7.2	6.9	24.0	23.1	22.7	24.1
YORKSHIRE AND THE HUMBER	**2,413.8**	**2,553.4**	**28.1**	**26.8**	**120.4**	**115.8**	**369.5**	**353.8**	**431.9**	**435.2**
East Riding of Yorkshire UA	**153.5**	**161.4**	**1.5**	**1.4**	**6.6**	**6.6**	**22.3**	**20.8**	**22.6**	**21.7**
Kingston upon Hull, City of UA	**119.1**	**124.3**	**1.5**	**1.3**	**6.3**	**5.8**	**19.2**	**18.5**	**23.5**	**24.2**
North East Lincolnshire UA	**76.7**	**81.3**	**0.9**	**0.9**	**4.0**	**3.8**	**12.9**	**12.3**	**12.2**	**12.7**
North Lincolnshire UA	**74.8**	**78.1**	**0.8**	**0.8**	**3.5**	**3.5**	**11.5**	**10.8**	**11.2**	**11.4**
York UA	**87.3**	**94.1**	**0.9**	**0.8**	**3.9**	**3.7**	**11.3**	**11.1**	**17.9**	**18.3**
North Yorkshire	**277.9**	**292.2**	**2.9**	**2.6**	**12.4**	**11.9**	**40.7**	**38.9**	**42.9**	**38.7**
Craven	25.8	27.9	0.3	0.3	1.1	1.1	3.8	3.7	3.5	3.3
Hambleton	41.5	42.7	0.4	0.3	1.8	1.8	6.0	5.7	5.8	5.1
Harrogate	73.3	78.2	0.8	0.7	3.4	3.2	10.6	10.4	11.8	11.0
Richmondshire	24.3	22.8	0.3	0.2	1.2	1.1	3.3	3.1	5.6	3.4
Ryedale	25.2	25.7	0.2	0.2	1.0	1.0	3.7	3.3	3.4	3.0
Scarborough	50.4	55.9	0.5	0.4	2.1	2.0	7.4	6.9	7.2	7.5
Selby	37.5	39.1	0.4	0.4	1.8	1.7	5.9	5.9	5.5	5.4
South Yorkshire (Met County)	**617.7**	**648.8**	**7.0**	**6.9**	**30.4**	**29.0**	**92.5**	**88.3**	**111.6**	**110.8**
Barnsley	106.2	112.0	1.2	1.1	5.2	4.9	16.2	15.6	17.0	17.1
Doncaster	140.1	146.8	1.6	1.6	6.9	6.7	21.9	21.1	23.2	22.6
Rotherham	120.8	127.6	1.5	1.4	6.1	6.0	19.1	18.1	19.4	20.0
Sheffield	250.6	262.5	2.8	2.7	12.2	11.4	35.2	33.6	52.1	51.0
West Yorkshire (Met County)	**1,006.9**	**1,073.3**	**12.5**	**12.1**	**53.4**	**51.5**	**159.0**	**153.0**	**189.9**	**197.4**
Bradford	225.4	242.5	3.3	3.2	13.5	13.3	38.5	37.5	43.0	45.7
Calderdale	93.1	99.4	1.1	1.1	5.1	4.7	14.8	14.2	14.7	15.1
Kirklees	189.1	199.9	2.3	2.4	10.5	10.2	30.0	28.6	34.0	35.1
Leeds	346.0	369.5	4.0	3.8	16.8	16.0	52.2	49.8	73.3	75.4
Wakefield	153.4	162.0	1.7	1.7	7.5	7.1	23.6	23.0	24.9	26.1

United Kingdom by countries and, within England, Government Office Regions; counties/unitary authorities, districts and London boroughs in Great Britain/district council areas in Northern Ireland

thousands

30-44		45-59		60-64		65-74		75 and over		Area
Males	Females	Males	Females	Males	Females	Males	Females	Males	Females	
275.5	**282.4**	**227.8**	**228.5**	**58.8**	**60.5**	**90.9**	**105.3**	**61.4**	**112.0**	**Greater Manchester (Met County)**
28.3	29.6	25.0	24.7	6.3	6.4	9.4	10.9	6.6	12.0	Bolton
20.6	21.1	17.7	17.8	4.3	4.6	6.6	7.7	4.4	8.0	Bury
41.3	42.3	28.5	27.9	7.5	8.0	12.5	14.4	8.9	16.1	Manchester
23.3	24.5	20.1	20.8	5.4	5.4	7.5	8.8	5.2	9.6	Oldham
22.3	23.2	19.4	19.7	4.7	4.6	7.3	8.6	4.9	8.6	Rochdale
24.1	23.2	18.9	18.7	5.1	5.4	8.7	9.8	5.7	10.9	Salford
32.3	33.5	28.1	28.2	7.3	7.5	11.5	13.6	7.8	14.2	Stockport
24.7	25.3	20.4	20.0	5.1	5.4	7.7	9.0	5.3	9.8	Tameside
24.5	25.2	19.6	20.2	4.9	5.3	8.3	9.6	6.0	10.2	Trafford
34.3	34.7	30.3	30.4	8.1	8.0	11.3	12.7	6.8	12.6	Wigan
120.7	**124.3**	**111.2**	**111.6**	**28.8**	**29.8**	**46.4**	**53.8**	**32.1**	**56.8**	**Lancashire**
9.5	10.0	8.5	8.5	2.1	2.1	3.2	3.8	2.4	4.1	Burnley
11.9	11.5	11.1	10.7	2.6	2.6	3.7	4.0	2.3	4.3	Chorley
7.6	7.7	7.5	7.4	2.0	2.3	3.8	4.6	3.0	5.3	Fylde
9.0	9.0	7.7	7.5	2.0	2.1	2.9	3.5	2.1	3.8	Hyndburn
13.3	13.8	12.0	12.3	3.2	3.4	5.5	6.7	4.2	7.5	Lancaster
9.2	9.6	8.8	8.5	2.1	2.0	3.2	3.9	2.4	4.3	Pendle
14.1	14.3	11.3	10.8	2.9	3.0	4.8	5.4	3.2	5.6	Preston
5.9	6.0	5.9	5.9	1.5	1.6	2.4	2.6	1.5	2.8	Ribble Valley
7.3	7.6	6.7	6.7	1.6	1.5	2.4	2.7	1.5	2.8	Rossendale
11.5	12.0	10.7	10.9	2.7	2.8	4.1	4.8	2.7	4.6	South Ribble
11.2	12.0	10.9	11.7	3.1	3.1	4.6	5.0	2.7	4.7	West Lancashire
10.2	10.9	10.1	10.7	3.2	3.5	5.7	6.8	4.0	7.0	Wyre
140.1	**155.2**	**122.9**	**129.2**	**33.5**	**36.6**	**57.5**	**68.1**	**35.3**	**66.3**	**Merseyside (Met County)**
16.2	18.9	12.5	13.3	3.6	3.8	6.1	7.6	3.0	5.7	Knowsley
45.7	49.4	36.2	38.0	10.1	10.6	17.5	20.1	10.3	19.1	Liverpool
19.0	19.9	17.4	17.8	4.7	5.0	7.4	8.2	4.2	7.9	St Helens
28.6	31.8	26.4	28.2	7.3	8.7	13.1	15.8	8.5	16.2	Sefton
30.6	35.2	30.4	31.8	7.8	8.6	13.5	16.4	9.3	17.5	Wirral
540.2	**555.3**	**469.7**	**473.0**	**121.0**	**126.6**	**196.2**	**227.9**	**136.9**	**239.0**	**YORKSHIRE AND THE HUMBER**
32.9	**33.6**	**33.7**	**34.7**	**9.0**	**9.3**	**14.9**	**16.5**	**9.9**	**16.8**	**East Riding of Yorkshire UA**
26.7	**26.5**	**21.2**	**20.4**	**5.4**	**5.6**	**9.2**	**11.0**	**6.2**	**10.9**	**Kingston upon Hull, City of UA**
16.8	**17.6**	**15.0**	**14.6**	**4.0**	**4.2**	**6.6**	**7.4**	**4.4**	**7.8**	**North East Lincolnshire UA**
16.9	**16.9**	**15.8**	**15.8**	**4.1**	**4.1**	**6.6**	**7.4**	**4.4**	**7.4**	**North Lincolnshire UA**
19.7	**20.3**	**16.5**	**17.2**	**4.3**	**4.6**	**7.3**	**8.6**	**5.4**	**9.4**	**York UA**
60.2	**63.1**	**59.3**	**60.5**	**15.7**	**16.4**	**25.5**	**28.5**	**18.4**	**31.4**	**North Yorkshire**
5.4	5.6	5.8	5.9	1.5	1.6	2.6	2.9	1.9	3.5	Craven
9.2	9.5	9.3	9.3	2.4	2.6	3.9	4.1	2.6	4.1	Hambleton
16.7	17.7	15.1	15.4	4.0	4.1	6.2	7.3	4.6	8.4	Harrogate
5.2	5.1	4.4	4.5	1.2	1.2	1.8	2.1	1.3	2.0	Richmondshire
5.1	5.2	5.6	5.6	1.6	1.6	2.7	2.9	2.0	3.0	Ryedale
9.8	10.7	10.9	11.5	3.1	3.4	5.3	6.1	4.1	7.2	Scarborough
8.9	9.3	8.2	8.2	1.9	1.8	2.9	3.2	2.0	3.2	Selby
140.7	**142.2**	**118.6**	**119.2**	**31.2**	**32.9**	**50.7**	**58.5**	**35.0**	**60.9**	**South Yorkshire (Met County)**
24.7	25.5	21.6	21.4	5.5	5.8	8.8	10.3	6.0	10.2	Barnsley
31.7	32.4	27.5	27.9	7.0	7.7	12.2	13.9	7.9	13.0	Doncaster
27.7	28.5	24.4	24.4	6.4	6.6	9.8	11.3	6.5	11.2	Rotherham
56.5	55.9	45.0	45.5	12.3	12.8	19.8	23.1	14.6	26.4	Sheffield
226.2	**235.0**	**189.6**	**190.5**	**47.3**	**49.4**	**75.5**	**90.0**	**53.3**	**94.4**	**West Yorkshire (Met County)**
49.1	51.2	40.4	40.7	10.2	10.6	16.1	20.0	11.3	20.4	Bradford
21.5	22.6	19.2	19.2	4.7	4.6	6.9	8.4	5.2	9.5	Calderdale
42.9	43.8	37.1	37.2	8.9	9.1	13.7	16.2	9.7	17.4	Kirklees
76.6	80.7	61.6	62.6	15.9	17.1	26.6	31.1	19.0	32.9	Leeds
36.2	36.8	31.4	30.8	7.7	8.1	12.2	14.2	8.2	14.2	Wakefield

Table 2.1 - *continued*

Area	All ages		Under 1		1 - 4		5 - 15		16 - 29	
	Males	Females	Males	Females	Males	Females	Males	Females	Males	Females
EAST MIDLANDS	**2,050.4**	**2,124.7**	**22.9**	**21.7**	**99.6**	**94.1**	**307.9**	**290.7**	**355.8**	**348.1**
Derby UA	**108.2**	**113.4**	**1.3**	**1.3**	**5.7**	**5.3**	**16.8**	**16.2**	**20.9**	**21.2**
Leicester UA	**134.7**	**145.0**	**2.0**	**1.9**	**7.8**	**7.4**	**21.9**	**21.3**	**30.6**	**33.5**
Nottingham UA	**132.5**	**134.4**	**1.6**	**1.5**	**6.3**	**6.0**	**19.2**	**18.3**	**35.5**	**33.4**
Rutland UA	**17.8**	**16.8**	**0.2**	**0.2**	**0.8**	**0.7**	**2.6**	**2.3**	**3.3**	**2.3**
Derbyshire	**361.0**	**373.9**	**3.8**	**3.6**	**17.1**	**16.1**	**53.8**	**50.0**	**54.7**	**53.7**
Amber Valley	57.1	59.5	0.6	0.6	2.6	2.7	8.3	7.9	8.6	8.5
Bolsover	35.3	36.6	0.4	0.3	1.7	1.6	5.3	4.9	5.4	5.5
Chesterfield	48.3	50.6	0.5	0.5	2.3	2.1	7.1	6.6	7.7	7.6
Derbyshire Dales	34.3	35.2	0.3	0.3	1.5	1.4	4.8	4.4	4.6	4.0
Erewash	53.9	56.2	0.6	0.6	2.6	2.5	8.3	8.0	8.6	8.7
High Peak	44.2	45.2	0.4	0.5	2.2	2.1	6.9	6.2	6.3	6.3
North East Derbyshire	47.5	49.4	0.4	0.4	2.0	1.9	6.8	6.4	7.1	6.7
South Derbyshire	40.5	41.2	0.5	0.4	2.1	1.9	6.2	5.6	6.3	6.4
Leicestershire	**301.6**	**308.7**	**3.3**	**3.1**	**14.4**	**13.4**	**43.7**	**41.5**	**51.5**	**48.2**
Blaby	44.9	45.5	0.5	0.5	2.2	2.1	6.6	6.3	7.4	6.6
Charnwood	76.4	77.2	0.8	0.8	3.4	3.1	10.8	10.2	16.6	14.7
Harborough	38.1	38.7	0.5	0.4	2.0	1.8	5.6	5.3	5.3	5.2
Hinckley and Bosworth	49.2	51.0	0.5	0.5	2.3	2.2	6.9	6.7	7.6	7.6
Melton	23.5	24.3	0.3	0.2	1.2	1.1	3.5	3.3	3.4	3.3
North West Leicestershire	42.3	43.4	0.5	0.4	2.1	2.0	6.1	5.7	6.4	6.4
Oadby and Wigston	27.2	28.6	0.3	0.2	1.3	1.1	4.2	4.0	4.8	4.6
Lincolnshire	**317.1**	**330.6**	**3.1**	**3.0**	**14.4**	**13.5**	**46.4**	**43.7**	**48.4**	**46.9**
Boston	27.4	28.4	0.3	0.3	1.2	1.2	3.8	3.6	4.3	4.0
East Lindsey	64.1	66.5	0.6	0.5	2.6	2.5	8.9	8.2	8.8	8.0
Lincoln	41.7	43.9	0.5	0.5	2.1	1.9	6.2	5.9	8.9	9.5
North Kesteven	46.2	48.2	0.4	0.5	2.2	2.0	6.7	6.5	6.6	6.3
South Holland	37.5	39.2	0.4	0.3	1.6	1.5	5.1	4.8	5.1	5.0
South Kesteven	61.2	63.7	0.6	0.6	2.9	2.9	9.7	8.9	9.3	9.1
West Lindsey	39.0	40.6	0.4	0.3	1.7	1.6	6.0	5.8	5.4	5.1
Northamptonshire	**311.2**	**319.3**	**3.8**	**3.6**	**16.0**	**15.3**	**49.0**	**45.7**	**53.2**	**51.7**
Corby	25.9	27.3	0.3	0.3	1.3	1.3	4.4	4.3	4.3	4.3
Daventry	36.1	36.0	0.4	0.4	1.8	1.7	5.8	5.3	5.5	4.9
East Northamptonshire	38.1	38.8	0.5	0.4	2.0	1.9	6.0	5.5	5.9	5.7
Kettering	40.3	41.8	0.5	0.5	2.0	1.9	6.2	5.8	6.7	6.6
Northampton	95.4	99.0	1.3	1.2	5.0	4.7	14.7	13.7	19.1	19.5
South Northamptonshire	39.6	39.9	0.4	0.5	2.0	1.9	6.3	5.9	5.6	5.1
Wellingborough	36.0	36.6	0.4	0.4	1.9	1.8	5.7	5.2	6.0	5.6
Nottinghamshire	**366.3**	**382.6**	**3.9**	**3.5**	**17.2**	**16.4**	**54.6**	**51.8**	**57.8**	**57.3**
Ashfield	54.5	57.1	0.6	0.6	2.8	2.6	8.1	7.9	8.7	9.2
Bassetlaw	53.3	54.6	0.6	0.5	2.5	2.5	8.0	7.5	8.3	7.8
Broxtowe	52.8	54.6	0.5	0.5	2.3	2.2	7.6	7.1	8.9	8.4
Gedling	54.3	57.5	0.6	0.5	2.4	2.3	7.9	7.5	8.3	8.5
Mansfield	47.7	50.3	0.5	0.4	2.2	2.2	7.6	7.2	7.7	7.9
Newark and Sherwood	51.9	54.5	0.6	0.5	2.5	2.3	7.8	7.5	7.7	7.6
Rushcliffe	51.9	53.9	0.5	0.5	2.5	2.3	7.6	7.1	8.2	8.0
WEST MIDLANDS	**2,575.5**	**2,691.6**	**30.4**	**29.4**	**132.8**	**125.4**	**397.5**	**379.6**	**453.3**	**450.5**
Herefordshire, County of UA	**85.4**	**89.5**	**0.8**	**0.8**	**4.0**	**3.8**	**12.6**	**11.9**	**11.9**	**11.6**
Stoke-on-Trent UA	**117.1**	**123.4**	**1.4**	**1.3**	**5.7**	**5.4**	**17.4**	**16.6**	**22.4**	**22.9**
Telford & Wrekin UA	**78.0**	**80.5**	**1.0**	**0.9**	**4.4**	**4.2**	**12.7**	**12.2**	**14.2**	**13.9**
Shropshire	**140.2**	**143.1**	**1.4**	**1.4**	**6.3**	**6.2**	**20.2**	**18.7**	**22.8**	**19.4**
Bridgnorth	26.7	25.8	0.2	0.2	1.0	1.1	3.4	3.3	5.3	3.4
North Shropshire	28.6	28.6	0.3	0.3	1.4	1.3	4.1	3.7	4.7	3.8
Oswestry	18.1	19.3	0.2	0.2	0.8	0.8	2.8	2.7	2.8	2.8
Shrewsbury and Atcham	47.1	48.8	0.5	0.5	2.2	2.1	7.2	6.4	7.6	7.1
South Shropshire	19.8	20.6	0.2	0.2	0.9	0.9	2.7	2.5	2.4	2.2

United Kingdom by countries and, within England, Government Office Regions; counties/unitary authorities, districts and London boroughs in Great Britain/district council areas in Northern Ireland,

thousands

30-44		45-59		60-64		65-74		75 and over		Area
Males	Females	Males	Females	Males	Females	Males	Females	Males	Females	
461.9	**471.6**	**410.2**	**410.4**	**102.9**	**104.8**	**169.1**	**187.3**	**120.0**	**195.8**	**EAST MIDLANDS**
24.0	**24.7**	**19.2**	**19.0**	**5.1**	**5.2**	**8.9**	**10.1**	**6.5**	**10.5**	**Derby UA**
29.8	**31.1**	**21.7**	**21.9**	**5.2**	**5.8**	**8.8**	**10.4**	**7.0**	**11.6**	**Leicester UA**
28.8	**28.5**	**19.9**	**19.1**	**5.1**	**5.4**	**9.3**	**10.8**	**6.9**	**11.5**	**Nottingham UA**
3.9	**3.5**	**3.6**	**3.7**	**1.0**	**1.0**	**1.5**	**1.6**	**1.0**	**1.7**	**Rutland UA**
82.7	**84.3**	**77.4**	**76.2**	**19.2**	**19.1**	**30.6**	**34.1**	**21.8**	**36.7**	**Derbyshire**
13.1	13.2	12.5	12.3	3.1	3.0	4.8	5.2	3.5	6.0	Amber Valley
8.2	8.3	7.1	7.0	1.9	1.8	3.1	3.6	2.2	3.6	Bolsover
11.0	11.3	9.9	9.9	2.4	2.5	4.1	4.8	3.2	5.4	Chesterfield
7.3	7.4	8.1	8.0	2.0	2.1	3.2	3.6	2.4	4.0	Derbyshire Dales
12.8	13.0	11.0	10.7	2.7	2.7	4.3	4.8	3.0	5.2	Erewash
10.5	10.7	9.5	9.2	2.3	2.2	3.4	3.8	2.4	4.2	High Peak
10.2	10.6	10.8	10.8	2.7	2.8	4.4	4.9	3.1	4.9	North East Derbyshire
9.6	9.9	8.5	8.4	2.0	2.0	3.1	3.3	2.1	3.3	South Derbyshire
68.6	**69.6**	**63.2**	**63.4**	**15.4**	**15.3**	**24.6**	**27.2**	**16.8**	**27.1**	**Leicestershire**
10.8	10.8	9.0	9.3	2.3	2.4	3.8	4.0	2.4	3.6	Blaby
16.2	16.7	15.1	15.1	3.6	3.5	5.9	6.5	4.0	6.6	Charnwood
9.0	9.1	8.5	8.3	2.0	2.0	3.1	3.3	2.2	3.4	Harborough
11.3	11.3	11.0	11.0	2.6	2.6	4.1	4.5	2.8	4.6	Hinckley and Bosworth
5.5	5.7	5.2	5.1	1.3	1.1	1.8	2.2	1.4	2.3	Melton
9.9	9.9	9.2	9.0	2.2	2.1	3.4	3.8	2.5	4.0	North West Leicestershire
5.8	6.2	5.3	5.4	1.4	1.5	2.5	2.9	1.6	2.6	Oadby and Wigston
66.7	**68.9**	**65.9**	**67.4**	**18.2**	**18.9**	**31.7**	**33.9**	**22.2**	**34.4**	**Lincolnshire**
5.5	5.7	5.8	5.8	1.6	1.7	2.8	3.0	1.9	3.1	Boston
12.0	12.6	13.8	14.4	4.3	4.6	7.9	8.1	5.4	7.8	East Lindsey
9.4	9.3	7.3	7.2	1.8	1.9	3.0	3.5	2.5	4.2	Lincoln
10.4	10.7	9.5	9.8	2.7	2.8	4.6	4.8	3.1	4.8	North Kesteven
7.7	7.7	7.8	8.2	2.4	2.5	4.5	4.8	3.0	4.4	South Holland
13.7	14.3	12.9	13.1	3.1	3.1	5.1	5.7	3.8	6.0	South Kesteven
8.1	8.6	8.7	8.9	2.3	2.3	3.9	4.0	2.6	4.0	West Lindsey
73.1	**74.5**	**63.0**	**63.1**	**14.6**	**14.4**	**22.4**	**24.7**	**16.1**	**26.3**	**Northamptonshire**
6.1	6.5	4.9	5.0	1.4	1.4	2.0	2.3	1.3	1.9	Corby
8.6	8.7	7.9	7.8	1.8	1.7	2.6	2.7	1.7	2.7	Daventry
8.9	8.9	8.1	8.0	1.9	1.8	2.8	3.1	2.1	3.5	East Northamptonshire
9.4	9.4	8.2	8.3	1.9	1.9	3.0	3.4	2.4	4.0	Kettering
22.5	22.9	17.7	17.7	4.0	4.0	6.4	7.3	4.9	8.0	Northampton
9.5	9.9	9.0	8.8	1.9	1.8	2.9	3.0	2.0	3.1	South Northamptonshire
8.2	8.3	7.3	7.4	1.8	1.8	2.7	3.0	1.9	3.1	Wellingborough
84.4	**86.5**	**76.3**	**76.7**	**19.3**	**19.7**	**31.1**	**34.6**	**21.6**	**36.0**	**Nottinghamshire**
12.9	12.9	11.1	11.0	2.9	2.9	4.4	4.8	3.0	5.3	Ashfield
12.3	12.3	11.2	11.2	2.9	2.9	4.6	5.0	3.0	5.0	Bassetlaw
12.3	12.5	10.8	11.1	2.8	2.8	4.5	5.0	3.1	5.1	Broxtowe
12.4	13.1	11.6	11.8	2.9	3.0	4.8	5.3	3.3	5.5	Gedling
10.8	11.4	9.5	9.5	2.4	2.5	4.0	4.7	2.9	4.5	Mansfield
11.5	11.9	11.1	11.2	2.9	2.9	4.6	5.2	3.2	5.4	Newark and Sherwood
12.2	12.4	11.0	11.0	2.5	2.6	4.3	4.7	3.0	5.1	Rushcliffe
570.3	**582.0**	**502.7**	**504.4**	**132.1**	**134.7**	**211.8**	**238.2**	**144.5**	**247.4**	**WEST MIDLANDS**
18.3	**18.5**	**18.4**	**18.5**	**4.9**	**5.1**	**8.3**	**9.3**	**6.1**	**9.9**	**Herefordshire, County of UA**
26.0	**26.3**	**22.5**	**21.8**	**5.6**	**5.9**	**9.3**	**11.3**	**6.7**	**11.8**	**Stoke-on-Trent UA**
18.3	**18.7**	**15.5**	**15.4**	**3.7**	**3.7**	**5.1**	**5.7**	**3.2**	**5.6**	**Telford & Wrekin UA**
30.2	**30.0**	**29.3**	**30.0**	**8.0**	**8.2**	**13.1**	**14.1**	**8.9**	**15.2**	**Shropshire**
5.4	5.3	5.8	5.9	1.6	1.6	2.4	2.5	1.5	2.5	Bridgnorth
6.3	6.1	5.9	5.9	1.6	1.6	2.6	2.8	1.8	3.0	North Shropshire
4.1	4.0	3.6	3.8	0.9	1.1	1.7	1.8	1.2	2.0	Oswestry
10.6	10.5	9.6	9.9	2.5	2.7	4.0	4.5	2.9	5.1	Shrewsbury and Atcham
3.9	4.0	4.4	4.5	1.3	1.3	2.4	2.4	1.6	2.5	South Shropshire

Table 2.1 - *continued*

Area	All ages		Under 1		1 - 4		5 - 15		16 - 29	
	Males	Females	Males	Females	Males	Females	Males	Females	Males	Females
Staffordshire	**396.4**	**410.7**	**4.3**	**4.0**	**18.4**	**17.9**	**58.0**	**56.2**	**65.1**	**62.4**
Cannock Chase	45.4	46.8	0.5	0.5	2.4	2.4	7.0	6.7	7.8	7.7
East Staffordshire	50.6	53.3	0.6	0.6	2.6	2.5	8.0	7.7	7.9	8.0
Lichfield	45.8	47.4	0.5	0.4	2.1	2.0	6.7	6.4	6.9	6.4
Newcastle-under-Lyme	59.3	62.8	0.6	0.6	2.7	2.5	8.3	8.0	10.7	11.1
South Staffordshire	52.3	53.6	0.5	0.5	2.2	2.2	7.5	7.4	7.9	7.1
Stafford	59.8	60.9	0.6	0.5	2.6	2.3	8.2	7.9	10.1	8.7
Staffordshire Moorlands	46.5	48.0	0.5	0.4	1.9	2.0	6.3	6.2	7.2	6.6
Tamworth	36.7	37.9	0.5	0.5	2.0	2.0	5.9	5.9	6.7	6.7
Warwickshire	**248.4**	**257.8**	**2.7**	**2.5**	**12.0**	**11.3**	**35.9**	**34.0**	**40.0**	**39.6**
North Warwickshire	30.4	31.4	0.3	0.3	1.5	1.4	4.5	4.3	4.6	4.7
Nuneaton and Bedworth	58.6	60.7	0.7	0.6	2.9	2.9	9.3	8.8	9.6	9.8
Rugby	43.4	44.1	0.5	0.5	2.1	2.0	6.5	5.9	6.9	6.7
Stratford-on-Avon	54.0	57.5	0.6	0.5	2.6	2.3	7.3	7.1	7.2	7.3
Warwick	62.0	64.1	0.6	0.6	2.9	2.7	8.3	7.9	11.8	11.1
West Midlands (Met County)	**1,243.9**	**1,310.4**	**16.0**	**15.6**	**69.0**	**65.0**	**201.9**	**193.3**	**235.0**	**239.4**
Birmingham	473.0	503.4	6.8	6.6	29.0	27.3	80.8	78.2	98.1	102.5
Coventry	149.1	151.7	1.8	1.8	7.7	7.4	23.3	21.7	32.7	30.5
Dudley	149.7	155.4	1.7	1.6	7.3	6.8	22.4	20.9	24.0	24.2
Sandwell	136.4	146.3	1.7	1.8	7.4	7.2	22.0	21.3	23.8	25.2
Solihull	96.7	102.8	1.0	1.0	4.8	4.3	15.7	15.0	14.2	14.2
Walsall	123.2	130.2	1.6	1.6	6.7	6.5	19.5	19.1	20.8	21.2
Wolverhampton	115.8	120.6	1.4	1.3	6.0	5.5	18.2	17.1	21.4	21.6
Worcestershire	**266.0**	**276.2**	**2.8**	**2.7**	**13.0**	**11.7**	**38.8**	**36.6**	**41.8**	**41.3**
Bromsgrove	43.2	44.7	0.4	0.4	2.0	1.8	6.4	6.1	6.1	5.7
Malvern Hills	35.1	37.1	0.3	0.3	1.4	1.3	5.0	4.8	4.8	4.5
Redditch	38.8	40.0	0.5	0.5	2.1	2.0	6.1	5.8	7.2	7.2
Worcester	45.5	47.8	0.6	0.6	2.6	2.3	6.5	6.2	8.1	9.1
Wychavon	55.8	57.3	0.5	0.5	2.6	2.4	7.9	7.4	7.8	7.6
Wyre Forest	47.6	49.2	0.4	0.4	2.3	1.9	6.8	6.3	7.7	7.2
EAST	**2,642.2**	**2,752.7**	**31.1**	**30.0**	**133.5**	**126.6**	**389.9**	**372.6**	**447.0**	**439.8**
Luton UA	**92.1**	**92.2**	**1.4**	**1.4**	**5.4**	**5.2**	**15.3**	**14.7**	**19.0**	**19.4**
Peterborough UA	**76.3**	**80.2**	**1.0**	**1.0**	**4.1**	**4.1**	**12.3**	**11.8**	**14.4**	**14.6**
Southend-on-Sea UA	**76.8**	**83.5**	**0.9**	**0.9**	**4.2**	**3.8**	**11.6**	**10.9**	**12.4**	**12.8**
Thurrock UA	**69.7**	**73.5**	**1.0**	**0.9**	**4.2**	**3.9**	**10.8**	**10.6**	**12.8**	**13.7**
Bedfordshire	**189.4**	**192.7**	**2.3**	**2.3**	**10.0**	**9.5**	**29.5**	**27.3**	**31.2**	**31.5**
Bedford	73.2	75.0	0.9	0.9	3.9	3.7	11.1	10.3	13.3	13.7
Mid Bedfordshire	60.6	60.7	0.8	0.8	3.3	3.1	9.2	8.7	9.4	9.0
South Bedfordshire	55.7	57.0	0.7	0.7	2.8	2.7	9.2	8.3	8.5	8.8
Cambridgeshire	**274.2**	**279.4**	**3.2**	**3.0**	**13.3**	**12.6**	**38.7**	**36.5**	**52.4**	**50.2**
Cambridge	54.3	54.5	0.6	0.5	2.0	2.0	5.7	5.2	18.4	17.0
East Cambridgeshire	36.3	37.1	0.4	0.4	1.8	1.7	5.3	5.0	5.6	5.5
Fenland	40.8	42.9	0.4	0.5	2.0	1.9	5.9	5.7	6.0	6.1
Huntingdonshire	78.1	79.0	1.0	0.9	4.3	4.0	12.4	11.6	12.0	12.1
South Cambridgeshire	64.6	65.8	0.7	0.7	3.2	3.1	9.5	9.1	10.4	9.5
Essex	**641.2**	**671.5**	**7.4**	**7.1**	**32.5**	**30.1**	**94.7**	**90.3**	**106.2**	**101.7**
Basildon	80.1	85.7	1.0	1.1	4.4	4.3	12.7	11.9	14.2	14.6
Braintree	65.2	67.3	0.8	0.8	3.5	3.2	9.8	9.5	10.5	10.0
Brentwood	33.2	35.2	0.3	0.3	1.5	1.5	4.9	4.5	5.1	4.7
Castle Point	42.4	44.3	0.4	0.4	1.9	1.8	6.2	6.0	6.7	6.3
Chelmsford	77.5	79.7	0.8	0.8	3.8	3.4	11.4	11.1	13.8	13.1
Colchester	77.3	78.7	0.9	0.9	4.0	3.7	11.0	10.5	15.8	14.2
Epping Forest	58.6	62.3	0.7	0.7	3.1	2.8	8.4	8.1	9.1	9.0
Harlow	38.3	40.6	0.6	0.5	2.2	2.0	6.0	5.7	7.1	7.4
Maldon	29.5	30.1	0.3	0.3	1.6	1.3	4.5	4.1	4.2	4.0
Rochford	38.2	40.5	0.4	0.4	1.9	1.8	5.6	5.5	5.7	5.6
Tendring	66.4	72.4	0.7	0.6	2.8	2.7	8.9	8.5	8.8	8.4
Uttlesford	34.4	34.6	0.4	0.3	1.7	1.6	5.2	4.9	5.1	4.5

United Kingdom by countries and, within England, Government Office Regions; counties/unitary authorities, districts and London boroughs in Great Britain/district council areas in Northern Ireland

thousands

30-44		45-59		60-64		65-74		75 and over		Area
Males	Females	Males	Females	Males	Females	Males	Females	Males	Females	
89.2	**90.8**	**85.2**	**85.2**	**22.0**	**21.8**	**33.1**	**36.7**	**21.1**	**35.7**	**Staffordshire**
11.1	10.9	9.0	9.0	2.2	2.3	3.4	3.8	2.1	3.4	Cannock Chase
11.9	12.2	10.2	10.1	2.6	2.6	4.1	4.9	2.7	4.6	East Staffordshire
10.0	10.2	10.6	10.8	2.9	2.8	3.9	4.1	2.4	4.2	Lichfield
13.0	13.2	12.2	12.1	3.2	3.2	5.1	6.0	3.5	6.1	Newcastle-under-Lyme
11.6	11.9	11.8	12.0	3.3	3.1	4.7	5.0	2.7	4.6	South Staffordshire
13.2	13.3	12.9	13.2	3.4	3.4	5.3	5.6	3.6	6.0	Stafford
10.0	10.2	10.7	10.6	2.8	2.7	4.3	4.8	2.9	4.6	Staffordshire Moorlands
8.5	8.9	7.8	7.4	1.6	1.6	2.3	2.5	1.3	2.3	Tamworth
56.8	**57.3**	**52.7**	**52.8**	**13.2**	**13.4**	**20.9**	**22.7**	**14.3**	**24.2**	**Warwickshire**
7.0	7.2	6.8	6.7	1.7	1.6	2.5	2.7	1.5	2.6	North Warwickshire
13.5	13.5	12.0	12.0	2.9	3.0	4.7	5.1	2.9	5.0	Nuneaton and Bedworth
10.0	9.9	8.8	8.9	2.3	2.2	3.6	3.7	2.6	4.2	Rugby
11.9	12.4	12.6	12.8	3.3	3.4	5.1	5.6	3.5	6.1	Stratford-on-Avon
14.2	14.3	12.5	12.4	3.0	3.1	5.0	5.6	3.8	6.4	Warwick
272.3	**279.9**	**221.1**	**222.9**	**60.5**	**62.3**	**99.6**	**113.9**	**68.5**	**118.2**	**West Midlands (Met County)**
101.7	106.6	76.6	78.2	20.6	21.4	34.5	39.3	24.9	43.2	Birmingham
32.5	31.8	25.0	25.0	6.6	7.2	11.1	12.5	8.4	14.0	Coventry
33.7	33.9	30.6	30.1	8.4	8.5	13.3	15.0	8.2	14.5	Dudley
30.7	31.9	24.5	24.4	7.2	7.0	11.3	13.6	7.8	14.0	Sandwell
21.3	22.4	20.4	21.2	5.1	5.2	8.4	9.9	5.7	9.6	Solihull
27.1	27.5	23.3	23.5	6.8	7.0	10.9	12.4	6.5	11.4	Walsall
25.3	25.9	20.6	20.6	5.8	5.9	10.1	11.4	7.1	11.4	Wolverhampton
59.3	**60.5**	**58.0**	**57.8**	**14.3**	**14.3**	**22.4**	**24.6**	**15.7**	**26.7**	**Worcestershire**
9.6	10.0	9.7	9.7	2.5	2.5	3.9	4.2	2.6	4.5	Bromsgrove
6.9	7.1	7.9	8.2	2.2	2.3	3.7	3.9	2.8	4.7	Malvern Hills
8.7	9.1	8.4	8.3	1.6	1.6	2.5	2.7	1.7	2.8	Redditch
11.6	11.3	8.5	8.3	2.0	2.1	3.2	3.8	2.3	4.1	Worcester
12.4	12.6	12.7	12.4	3.2	3.2	5.1	5.5	3.6	5.7	Wychavon
10.1	10.4	10.7	10.8	2.9	2.7	4.0	4.5	2.7	4.9	Wyre Forest
603.0	**607.4**	**524.7**	**533.7**	**131.6**	**135.0**	**221.4**	**244.6**	**159.9**	**262.9**	**EAST**
21.4	**20.9**	**15.2**	**15.0**	**4.1**	**3.8**	**6.5**	**6.3**	**3.7**	**5.7**	**Luton UA**
17.5	**18.2**	**14.0**	**14.4**	**3.4**	**3.4**	**5.6**	**6.4**	**3.9**	**6.2**	**Peterborough UA**
17.4	**17.7**	**14.5**	**15.1**	**3.7**	**3.8**	**6.5**	**7.8**	**5.7**	**10.7**	**Southend-on-Sea UA**
17.1	**17.0**	**13.4**	**13.3**	**2.9**	**3.1**	**4.5**	**5.4**	**3.2**	**5.6**	**Thurrock UA**
46.2	**45.4**	**37.9**	**37.4**	**9.0**	**8.8**	**14.0**	**15.3**	**9.3**	**15.2**	**Bedfordshire**
17.0	16.4	14.1	13.9	3.4	3.4	5.5	6.2	4.0	6.4	Bedford
15.6	15.3	12.4	12.2	2.9	2.7	4.2	4.5	2.8	4.5	Mid Bedfordshire
13.6	13.8	11.3	11.2	2.7	2.7	4.2	4.6	2.6	4.3	South Bedfordshire
63.5	**63.1**	**54.4**	**54.5**	**13.1**	**13.0**	**20.7**	**22.4**	**14.8**	**24.1**	**Cambridgeshire**
11.8	10.9	8.0	8.4	1.9	2.0	3.2	3.7	2.7	4.8	Cambridge
8.4	8.6	7.5	7.4	1.9	1.8	3.1	3.4	2.2	3.4	East Cambridgeshire
8.9	9.0	8.4	8.4	2.2	2.3	4.1	4.5	2.8	4.6	Fenland
19.3	19.1	16.5	16.2	3.8	3.7	5.3	5.7	3.6	5.8	Huntingdonshire
15.1	15.5	14.0	14.2	3.2	3.1	4.9	5.2	3.6	5.5	South Cambridgeshire
143.8	**146.8**	**130.7**	**135.1**	**32.5**	**34.3**	**54.7**	**60.8**	**38.8**	**65.2**	**Essex**
18.7	19.4	15.2	16.2	3.7	4.1	6.3	7.3	3.9	7.0	Basildon
15.4	15.5	13.7	13.8	3.1	3.0	4.8	5.3	3.6	6.2	Braintree
7.3	7.8	6.9	7.2	1.8	1.9	3.1	3.6	2.2	3.7	Brentwood
8.7	9.0	9.5	9.9	2.4	2.6	3.9	4.2	2.6	4.1	Castle Point
18.0	18.2	16.1	16.1	3.7	3.8	6.0	6.5	4.0	6.7	Chelmsford
17.5	17.2	14.8	15.4	3.5	3.7	5.7	6.2	4.0	7.1	Colchester
13.5	14.2	12.3	12.7	3.0	3.1	5.0	5.5	3.7	6.2	Epping Forest
9.5	9.6	6.6	6.8	1.5	1.8	3.0	3.7	1.9	3.0	Harlow
6.5	6.8	6.7	6.8	1.6	1.6	2.5	2.5	1.5	2.7	Maldon
8.4	8.7	8.1	8.4	2.1	2.2	3.6	3.9	2.4	3.9	Rochford
12.4	12.5	13.2	14.1	4.3	4.7	8.2	9.3	7.1	11.6	Tendring
7.8	8.0	7.7	7.6	1.8	1.7	2.8	2.8	1.9	3.1	Uttlesford

Table 2.1 - *continued*

Area	All ages		Under 1		1 - 4		5 - 15		16 - 29	
	Males	Females	Males	Females	Males	Females	Males	Females	Males	Females
Hertfordshire	**505.6**	**529.3**	**6.4**	**6.3**	**26.7**	**25.6**	**76.0**	**73.9**	**84.3**	**85.5**
Broxbourne	42.3	44.9	0.5	0.5	2.2	2.2	6.5	6.3	7.3	7.4
Dacorum	67.8	70.0	0.9	0.9	3.6	3.3	10.3	9.9	10.9	10.7
East Hertfordshire	63.3	65.8	0.8	0.8	3.4	3.2	9.5	9.3	10.2	10.5
Hertsmere	45.6	48.9	0.6	0.6	2.5	2.2	7.0	6.7	7.3	7.7
North Hertfordshire	57.1	60.0	0.7	0.7	2.9	2.9	8.5	8.4	8.9	8.9
St Albans	63.5	65.6	0.9	0.9	3.5	3.4	9.1	8.9	10.1	9.9
Stevenage	39.2	40.6	0.5	0.5	2.2	2.0	6.6	6.1	6.9	7.1
Three Rivers	40.1	42.8	0.5	0.5	2.1	2.0	6.0	6.0	6.1	6.2
Watford	39.2	40.5	0.5	0.5	2.0	2.0	5.7	5.6	7.6	7.9
Welwyn Hatfield	47.4	50.2	0.5	0.5	2.3	2.3	6.8	6.7	9.1	9.1
Norfolk	**388.5**	**409.4**	**3.9**	**3.7**	**17.1**	**16.5**	**52.7**	**50.7**	**61.2**	**60.7**
Breckland	60.2	61.4	0.6	0.6	2.9	2.7	8.6	8.2	9.2	8.7
Broadland	57.9	60.9	0.6	0.6	2.5	2.5	7.9	7.4	7.9	8.0
Great Yarmouth	44.1	46.9	0.4	0.4	2.0	2.0	6.4	6.1	6.7	6.8
Kings Lynn and West Norfolk	65.9	69.6	0.7	0.7	2.8	2.8	9.0	8.6	9.4	9.4
North Norfolk	47.5	51.0	0.4	0.3	1.8	1.8	6.0	5.7	6.2	5.9
Norwich	58.9	62.7	0.7	0.6	2.6	2.5	7.4	7.3	14.4	14.8
South Norfolk	53.9	56.9	0.5	0.5	2.5	2.3	7.4	7.4	7.3	7.1
Suffolk	**328.4**	**341.0**	**3.6**	**3.3**	**16.1**	**15.3**	**48.3**	**46.0**	**53.0**	**49.8**
Babergh	40.8	42.8	0.4	0.4	1.9	1.9	5.9	5.7	6.0	5.6
Forest Heath	27.9	27.7	0.4	0.3	1.6	1.5	3.9	3.9	5.7	5.1
Ipswich	57.5	59.7	0.7	0.7	3.0	2.9	8.9	8.5	11.1	10.8
Mid Suffolk	43.4	43.6	0.5	0.4	2.1	2.0	6.3	5.9	6.5	5.8
St Edmundsbury	48.7	49.6	0.5	0.5	2.4	2.2	6.9	6.5	8.4	7.4
Suffolk Coastal	55.9	59.4	0.5	0.5	2.6	2.5	8.3	7.9	7.4	7.1
Waveney	54.2	58.3	0.5	0.5	2.6	2.4	8.1	7.7	7.8	7.9
LONDON	**3,479.5**	**3,708.5**	**49.6**	**47.4**	**194.1**	**186.9**	**496.7**	**475.8**	**757.5**	**800.8**
Inner London	**1,344.6**	**1,427.1**	**20.7**	**19.8**	**76.1**	**73.6**	**177.2**	**172.6**	**334.4**	**366.2**
Camden	95.7	102.8	1.3	1.3	4.7	4.5	10.8	10.5	25.2	28.7
City of London	3.8	3.4	0.0	0.0	0.1	0.1	0.2	0.2	0.8	0.8
Hackney	97.3	106.0	1.8	1.7	6.8	6.5	15.4	15.4	21.3	25.4
Hammersmith and Fulham	79.2	86.3	1.1	1.0	4.2	3.9	8.4	8.7	21.0	23.9
Haringey	104.0	112.8	1.6	1.5	5.8	5.8	15.0	14.8	25.4	26.7
Islington	84.4	91.7	1.2	1.2	4.5	4.2	10.7	10.5	20.2	23.8
Kensington and Chelsea	76.1	83.1	1.1	1.1	3.9	3.8	7.7	7.2	15.7	18.5
Lambeth	131.6	135.2	2.0	1.9	7.3	6.9	16.8	16.3	34.6	35.6
Lewisham	120.3	129.1	1.9	1.8	7.2	6.9	17.5	17.4	26.3	27.7
Newham	120.2	124.1	2.2	2.1	8.3	8.1	22.2	20.7	28.8	30.0
Southwark	120.2	125.2	1.8	1.8	6.9	6.9	16.5	15.9	29.2	29.9
Tower Hamlets	98.5	98.1	1.6	1.5	6.1	6.0	15.1	14.7	27.9	30.1
Wandsworth	124.1	136.8	2.0	1.8	6.6	6.3	13.2	12.8	34.7	40.3
Westminster	89.1	92.6	1.1	1.1	3.7	3.6	7.7	7.5	23.3	24.6
Outer London	**2,134.9**	**2,281.4**	**28.9**	**27.6**	**118.0**	**113.3**	**319.5**	**303.2**	**423.1**	**434.6**
Barking and Dagenham	78.3	86.1	1.1	1.1	5.1	5.2	13.4	12.5	14.6	16.6
Barnet	150.3	165.0	2.0	2.0	8.3	8.0	22.3	21.4	30.6	32.8
Bexley	105.4	113.4	1.3	1.3	5.5	5.3	16.6	16.1	17.6	18.0
Brent	128.1	135.7	1.7	1.6	6.5	6.5	18.1	17.7	30.9	31.3
Bromley	142.1	154.0	1.9	1.8	7.7	7.2	20.4	19.9	22.9	23.1
Croydon	159.7	171.8	2.2	2.1	9.3	9.0	25.9	24.1	29.3	31.0
Ealing	148.0	153.5	2.1	2.0	7.7	7.6	20.8	19.6	33.8	33.9
Enfield	131.2	143.2	1.9	1.8	7.3	7.2	20.3	19.6	25.1	27.0
Greenwich	103.3	112.0	1.7	1.5	6.3	6.1	16.3	14.9	21.5	23.4
Harrow	100.6	107.3	1.2	1.1	5.0	4.7	15.4	14.2	20.2	19.9
Havering	108.2	116.5	1.2	1.1	5.2	4.9	16.5	15.6	18.0	17.7
Hillingdon	117.6	125.5	1.6	1.5	6.8	6.2	18.1	17.3	22.1	23.7
Hounslow	104.5	108.1	1.5	1.4	5.8	5.5	15.0	14.5	23.2	23.5
Kingston upon Thames	72.2	75.4	0.9	0.9	3.8	3.6	9.5	9.2	16.3	15.5
Merton	91.8	96.5	1.4	1.3	5.2	4.9	12.4	11.5	19.8	19.5

United Kingdom by countries and, within England, Government Office Regions; counties/unitary authorities, districts and London boroughs in Great Britain/district council areas in Northern Ireland

thousands

30-44		45-59		60-64		65-74		75 and over		Area
Males	Females	Males	Females	Males	Females	Males	Females	Males	Females	
123.8	**125.4**	**98.8**	**98.6**	**22.9**	**23.9**	**39.1**	**44.0**	**27.5**	**46.1**	**Hertfordshire**
10.0	10.4	8.2	8.6	2.1	2.2	3.6	3.9	2.0	3.4	Broxbourne
16.6	16.8	13.6	13.4	3.0	3.1	5.2	5.9	3.7	6.0	Dacorum
15.9	16.2	13.0	12.9	2.9	2.9	4.7	5.0	3.0	5.1	East Hertfordshire
10.7	11.4	9.2	9.1	2.0	2.1	3.4	4.0	3.0	5.0	Hertsmere
14.0	14.0	11.4	11.4	2.7	2.9	4.6	4.9	3.3	5.8	North Hertfordshire
16.0	16.0	12.7	12.6	3.0	3.0	4.9	5.4	3.4	5.6	St Albans
10.0	10.1	6.7	6.7	1.6	1.7	2.8	3.4	1.8	2.9	Stevenage
9.3	9.7	8.5	8.4	1.9	2.0	3.3	3.8	2.5	4.2	Three Rivers
10.6	9.9	6.8	6.7	1.6	1.6	2.5	2.9	1.8	3.3	Watford
10.5	10.9	8.8	8.7	2.1	2.3	4.1	4.8	3.0	4.8	Welwyn Hatfield
81.2	**81.4**	**80.1**	**82.8**	**22.6**	**23.2**	**39.6**	**43.4**	**30.1**	**47.1**	**Norfolk**
12.7	12.3	12.4	12.5	3.4	3.6	6.1	6.1	4.4	6.9	Breckland
13.0	13.0	12.4	12.9	3.5	3.6	5.8	6.4	4.3	6.5	Broadland
9.0	9.3	9.4	9.4	2.6	2.6	4.3	5.0	3.3	5.4	Great Yarmouth
13.7	13.6	13.4	14.2	4.0	4.2	7.4	8.0	5.4	8.1	Kings Lynn and West Norfolk
8.6	8.9	10.4	10.9	3.2	3.4	6.3	6.7	4.8	7.3	North Norfolk
13.1	12.6	10.0	10.3	2.5	2.6	4.5	5.3	3.8	6.8	Norwich
11.1	11.9	12.1	12.5	3.3	3.2	5.4	5.8	4.2	6.0	South Norfolk
71.0	**71.5**	**65.8**	**67.6**	**17.5**	**17.7**	**30.1**	**32.7**	**23.0**	**37.0**	**Suffolk**
8.4	8.8	9.1	9.4	2.3	2.3	3.9	4.2	2.8	4.5	Babergh
6.9	6.2	4.6	4.8	1.2	1.2	2.1	2.2	1.5	2.5	Forest Heath
12.8	12.7	10.2	10.2	2.6	2.6	4.7	5.2	3.5	6.2	Ipswich
9.4	9.6	9.3	9.2	2.4	2.3	4.0	4.0	2.9	4.4	Mid Suffolk
11.2	10.9	9.7	10.4	2.7	2.7	4.0	4.3	2.9	4.7	St Edmundsbury
11.6	12.1	11.9	12.2	3.2	3.3	5.8	6.4	4.6	7.3	Suffolk Coastal
10.8	11.2	10.9	11.4	3.1	3.3	5.6	6.4	4.8	7.5	Waveney
909.5	**939.7**	**561.1**	**594.9**	**136.8**	**145.4**	**219.0**	**249.1**	**155.1**	**268.7**	**LONDON**
379.3	**382.3**	**188.2**	**200.5**	**47.0**	**49.7**	**72.8**	**80.5**	**48.8**	**81.9**	**Inner London**
26.8	26.7	14.4	15.4	3.3	3.5	5.3	5.9	3.9	6.2	Camden
1.1	0.9	0.9	0.7	0.2	0.1	0.2	0.3	0.2	0.3	City of London
26.9	29.1	13.5	14.1	3.3	3.3	5.2	5.2	3.1	5.5	Hackney
22.8	23.5	11.5	12.0	2.9	3.2	4.3	4.8	3.0	5.3	Hammersmith and Fulham
28.7	31.2	14.7	16.5	3.7	4.1	5.8	6.2	3.3	5.9	Haringey
24.9	25.5	12.2	13.0	3.0	3.1	4.7	5.3	3.0	4.9	Islington
22.3	22.4	13.7	15.3	3.3	3.7	4.7	5.5	3.7	5.6	Kensington and Chelsea
39.5	38.7	16.6	17.7	4.1	4.4	6.5	6.9	4.3	6.8	Lambeth
34.6	36.0	17.6	18.6	4.3	4.5	6.3	7.8	4.8	8.4	Lewisham
29.6	30.4	15.8	16.3	3.9	4.1	5.7	6.3	3.7	6.1	Newham
34.9	34.7	16.2	17.0	4.0	4.3	6.4	7.2	4.3	7.4	Southwark
25.2	21.6	10.9	11.3	3.2	3.1	5.4	5.2	3.1	4.7	Tower Hamlets
36.7	37.5	15.7	17.4	4.2	4.4	6.4	7.5	4.6	8.6	Wandsworth
25.4	24.2	14.4	15.2	3.6	3.7	5.7	6.4	4.0	6.3	Westminster
530.2	**557.3**	**372.9**	**394.3**	**89.8**	**95.7**	**146.2**	**168.6**	**106.3**	**186.8**	**Outer London**
18.7	20.0	12.7	13.0	3.0	3.2	5.2	6.8	4.3	7.7	Barking and Dagenham
36.3	39.0	26.2	28.3	5.9	6.8	10.6	12.2	8.1	14.6	Barnet
25.2	26.1	19.9	21.1	5.0	5.4	8.5	9.8	6.0	10.3	Bexley
32.3	34.8	19.7	21.2	5.4	5.7	8.7	9.0	4.8	7.8	Brent
34.7	36.1	27.5	29.1	6.6	7.3	11.6	13.9	8.8	15.6	Bromley
39.5	43.8	28.7	30.1	6.6	7.2	10.5	12.0	7.7	12.5	Croydon
38.8	39.4	23.9	25.2	6.1	6.0	8.7	9.7	6.3	10.2	Ealing
32.7	34.9	22.5	24.4	5.8	6.2	9.3	10.4	6.4	11.7	Enfield
25.8	27.7	16.8	17.2	4.0	4.4	6.1	7.7	4.8	9.1	Greenwich
23.4	25.2	18.4	19.8	4.5	4.9	7.1	8.2	5.4	9.3	Harrow
24.3	25.3	21.4	22.8	5.3	5.9	9.6	11.6	6.9	11.7	Havering
28.9	30.2	21.0	21.4	5.0	5.5	8.4	9.6	5.8	10.1	Hillingdon
27.1	27.2	17.1	18.0	4.2	4.2	6.3	6.8	4.2	7.0	Hounslow
18.1	18.2	13.0	13.5	2.7	2.8	4.3	5.0	3.6	6.7	Kingston upon Thames
24.5	25.1	15.0	16.3	3.6	3.7	5.7	6.7	4.2	7.6	Merton

Table 2.1 - *continued*

Area	All ages		Under 1		1 - 4		5 - 15		16 - 29	
	Males	Females	Males	Females	Males	Females	Males	Females	Males	Females
Redbridge	116.3	123.0	1.5	1.4	6.6	6.2	18.0	17.3	23.0	22.9
Richmond upon Thames	83.6	89.2	1.3	1.2	4.8	4.6	10.6	10.1	14.3	14.9
Sutton	87.1	93.1	1.1	1.1	4.8	4.6	13.6	12.5	15.7	16.1
Waltham Forest	106.6	112.1	1.5	1.5	6.5	6.0	16.2	15.1	24.3	23.8
SOUTH EAST	**3,909.6**	**4,097.3**	**45.8**	**43.7**	**196.0**	**186.1**	**577.4**	**545.4**	**678.9**	**664.8**
Bracknell Forest UA	**54.9**	**54.7**	**0.8**	**0.7**	**3.2**	**3.0**	**8.7**	**8.0**	**10.2**	**9.5**
Brighton and Hove UA	**120.1**	**128.0**	**1.4**	**1.3**	**5.3**	**5.1**	**14.1**	**14.0**	**26.0**	**27.8**
Isle of Wight UA	**63.8**	**69.1**	**0.6**	**0.6**	**2.7**	**2.5**	**9.1**	**8.6**	**9.0**	**8.5**
Medway UA	**123.0**	**126.7**	**1.6**	**1.4**	**6.9**	**6.3**	**20.4**	**19.4**	**22.4**	**22.4**
Milton Keynes UA	**103.2**	**104.4**	**1.4**	**1.4**	**5.9**	**5.7**	**17.2**	**15.9**	**19.6**	**19.3**
Portsmouth UA	**92.2**	**94.7**	**1.2**	**1.1**	**4.5**	**4.3**	**12.9**	**12.3**	**21.2**	**20.1**
Reading UA	**72.2**	**71.0**	**1.0**	**0.9**	**3.6**	**3.5**	**9.5**	**9.1**	**18.7**	**17.4**
Slough UA	**59.3**	**59.7**	**0.9**	**0.8**	**3.3**	**3.2**	**9.6**	**8.7**	**12.6**	**13.3**
Southampton UA	**108.8**	**108.8**	**1.2**	**1.2**	**5.0**	**4.6**	**14.1**	**13.7**	**30.3**	**28.1**
West Berkshire UA	**71.7**	**72.7**	**0.9**	**0.8**	**3.6**	**3.5**	**10.9**	**10.4**	**11.8**	**11.6**
Windsor and Maidenhead UA	**65.9**	**67.7**	**0.8**	**0.8**	**3.4**	**3.2**	**9.7**	**8.6**	**11.3**	**10.3**
Wokingham UA	**75.2**	**75.2**	**0.9**	**0.9**	**3.8**	**3.6**	**11.6**	**10.8**	**13.5**	**12.2**
Buckinghamshire	**234.9**	**244.3**	**3.1**	**2.9**	**12.6**	**11.7**	**35.7**	**34.1**	**38.2**	**38.3**
Aylesbury Vale	82.4	83.5	1.2	1.1	4.5	4.2	12.9	11.7	13.9	13.6
Chiltern	43.1	46.1	0.5	0.5	2.3	2.1	6.5	6.4	6.0	6.0
South Bucks	30.0	31.9	0.3	0.3	1.5	1.4	4.6	4.3	4.0	4.2
Wycombe	79.3	82.8	1.1	1.0	4.4	4.1	11.7	11.7	14.2	14.6
East Sussex	**233.6**	**259.6**	**2.4**	**2.4**	**11.0**	**10.6**	**34.3**	**32.3**	**32.0**	**32.7**
Eastbourne	41.8	48.1	0.4	0.5	1.9	1.9	5.9	5.5	6.5	7.2
Hastings	40.9	44.5	0.5	0.5	2.2	2.2	6.4	6.2	6.5	6.9
Lewes	44.1	48.2	0.5	0.4	2.1	1.9	6.6	5.9	5.8	5.9
Rother	39.9	45.5	0.3	0.3	1.6	1.6	5.5	5.1	4.8	4.5
Wealden	66.8	73.3	0.7	0.7	3.2	3.0	10.0	9.5	8.3	8.3
Hampshire	**608.6**	**632.2**	**6.9**	**6.6**	**30.0**	**28.5**	**91.2**	**86.3**	**100.4**	**93.6**
Basingstoke and Deane	75.6	77.0	0.9	0.9	4.1	4.0	11.4	10.9	13.0	12.5
East Hampshire	53.8	55.6	0.6	0.5	2.7	2.5	8.2	7.8	8.7	7.3
Eastleigh	57.1	59.2	0.7	0.6	2.9	2.7	9.0	8.6	9.4	9.2
Fareham	53.0	55.1	0.6	0.5	2.5	2.3	7.9	7.3	7.9	7.3
Gosport	37.4	39.0	0.5	0.4	1.9	1.8	5.6	5.4	7.2	6.8
Hart	42.1	41.5	0.5	0.4	2.1	2.0	6.2	5.8	7.5	6.0
Havant	56.1	60.7	0.6	0.6	2.7	2.5	8.6	8.3	8.4	8.6
New Forest	81.1	88.4	0.8	0.8	3.5	3.3	11.5	10.9	10.8	10.4
Rushmoor	45.8	45.0	0.7	0.6	2.6	2.5	6.8	6.4	10.0	9.0
Test Valley	54.0	56.0	0.6	0.6	2.7	2.6	8.6	8.1	8.2	7.8
Winchester	52.6	54.7	0.6	0.5	2.4	2.2	7.5	6.8	9.4	8.9
Kent	**644.8**	**686.3**	**7.2**	**7.0**	**32.8**	**31.1**	**99.2**	**94.5**	**104.2**	**105.0**
Ashford	50.1	52.9	0.7	0.6	2.7	2.5	7.8	7.6	7.7	7.8
Canterbury	64.2	71.2	0.6	0.6	2.9	2.8	9.3	8.8	12.9	13.9
Dartford	42.2	43.8	0.5	0.5	2.3	2.2	6.5	6.2	7.4	7.6
Dover	50.3	54.3	0.5	0.5	2.5	2.3	7.7	7.5	7.8	7.7
Gravesham	46.9	48.9	0.6	0.5	2.5	2.4	7.6	7.1	7.6	7.9
Maidstone	68.5	70.7	0.8	0.7	3.3	3.2	10.2	9.2	11.4	11.0
Sevenoaks	52.9	56.4	0.6	0.6	2.7	2.5	8.0	7.8	7.5	7.3
Shepway	46.1	50.2	0.5	0.5	2.3	2.2	6.8	6.7	6.9	7.0
Swale	60.7	62.4	0.7	0.7	3.2	3.1	9.6	9.2	10.1	9.8
Thanet	60.0	66.7	0.6	0.6	3.0	2.8	9.6	8.9	8.9	9.4
Tonbridge and Malling	52.7	55.0	0.6	0.6	2.8	2.7	8.5	8.0	7.9	7.7
Tunbridge Wells	50.2	53.9	0.6	0.6	2.6	2.6	7.7	7.5	8.0	8.0
Oxfordshire	**299.5**	**306.4**	**3.5**	**3.4**	**14.7**	**14.2**	**42.7**	**39.9**	**59.1**	**57.1**
Cherwell	65.3	66.7	0.9	0.8	3.6	3.5	9.6	9.4	11.2	11.0
Oxford	66.3	67.9	0.7	0.7	2.6	2.6	7.8	7.1	21.5	21.6
South Oxfordshire	63.3	65.0	0.8	0.8	3.3	3.1	9.3	8.7	9.7	9.2
Vale of White Horse	57.6	58.2	0.6	0.6	2.8	2.7	8.9	8.0	9.5	8.6
West Oxfordshire	47.1	48.6	0.5	0.5	2.4	2.2	7.1	6.7	7.2	6.7

United Kingdom by countries and, within England, Government Office Regions; counties/unitary authorities, districts and London boroughs in Great Britain/district council areas in Northern Ireland

thousands

30-44		45-59		60-64		65-74		75 and over		Area
Males	Females	Males	Females	Males	Females	Males	Females	Males	Females	
27.2	28.7	21.1	22.0	4.9	5.0	8.0	9.1	5.9	10.5	Redbridge
23.4	23.9	16.4	17.1	3.4	3.4	5.2	6.1	4.4	8.0	Richmond upon Thames
21.9	22.8	15.9	16.4	3.6	3.9	6.0	7.2	4.5	8.4	Sutton
27.4	29.1	15.9	17.5	4.2	4.2	6.3	6.8	4.2	8.1	Waltham Forest
897.6	**912.9**	**775.4**	**786.1**	**190.0**	**196.2**	**313.4**	**355.5**	**235.1**	**406.6**	**SOUTH EAST**
15.0	**14.7**	**10.1**	**9.7**	**2.0**	**2.1**	**3.0**	**3.4**	**2.0**	**3.5**	**Bracknell Forest UA**
31.4	**29.8**	**20.5**	**20.5**	**5.4**	**5.2**	**8.6**	**10.4**	**7.5**	**13.9**	**Brighton and Hove UA**
12.4	**12.9**	**13.8**	**14.4**	**4.0**	**4.0**	**6.6**	**7.9**	**5.5**	**9.7**	**Isle of Wight UA**
29.2	**29.6**	**23.6**	**23.8**	**5.6**	**5.6**	**8.3**	**9.1**	**5.0**	**9.2**	**Medway UA**
25.8	**26.0**	**20.5**	**20.0**	**3.9**	**3.8**	**5.3**	**6.1**	**3.7**	**6.3**	**Milton Keynes UA**
21.6	**21.2**	**15.4**	**14.9**	**3.8**	**3.9**	**6.2**	**7.5**	**5.4**	**9.5**	**Portsmouth UA**
17.9	**16.2**	**11.3**	**10.9**	**2.7**	**2.7**	**4.4**	**4.8**	**3.2**	**5.6**	**Reading UA**
14.9	**14.8**	**9.8**	**9.0**	**2.2**	**2.0**	**3.6**	**4.0**	**2.4**	**3.8**	**Slough UA**
22.8	**21.6**	**18.0**	**16.8**	**4.3**	**4.2**	**7.2**	**8.3**	**5.8**	**10.3**	**Southampton UA**
17.8	**17.5**	**15.1**	**14.9**	**3.4**	**3.2**	**5.0**	**5.4**	**3.3**	**5.5**	**West Berkshire UA**
15.5	**15.9**	**13.2**	**13.7**	**3.3**	**3.3**	**5.2**	**5.7**	**3.5**	**6.1**	**Windsor and Maidenhead UA**
18.4	**18.5**	**15.6**	**15.7**	**3.5**	**3.5**	**5.1**	**5.3**	**2.9**	**4.8**	**Wokingham UA**
54.3	**56.3**	**48.9**	**49.6**	**11.8**	**11.7**	**18.3**	**19.5**	**12.0**	**20.0**	**Buckinghamshire**
20.3	20.7	16.7	16.6	3.7	3.5	5.7	5.9	3.6	6.2	Aylesbury Vale
9.1	10.0	9.8	10.0	2.4	2.5	3.9	4.2	2.6	4.3	Chiltern
6.5	7.0	6.6	6.8	1.8	1.8	2.8	3.0	1.9	3.1	South Bucks
18.3	18.7	15.8	16.2	3.9	3.9	5.9	6.3	3.9	6.4	Wycombe
47.4	**50.0**	**48.0**	**50.5**	**12.9**	**14.3**	**24.1**	**28.8**	**21.6**	**37.9**	**East Sussex**
8.7	8.9	7.6	8.1	2.2	2.4	4.2	5.5	4.4	8.1	Eastbourne
9.1	9.2	8.3	8.2	1.9	2.1	3.3	4.0	2.7	5.3	Hastings
8.9	9.4	9.2	9.7	2.4	2.7	4.6	5.4	4.0	6.7	Lewes
7.0	7.7	8.2	9.0	2.5	2.9	5.0	6.1	5.0	8.3	Rother
13.8	14.9	14.6	15.6	3.9	4.2	6.9	7.7	5.6	9.4	Wealden
139.5	**143.0**	**124.5**	**126.9**	**30.4**	**31.4**	**50.1**	**55.8**	**35.5**	**60.1**	**Hampshire**
19.0	19.0	15.5	15.5	3.5	3.4	5.1	5.5	3.1	5.4	Basingstoke and Deane
12.0	12.7	11.6	11.8	2.7	2.9	4.3	4.7	3.1	5.3	East Hampshire
13.4	13.8	11.8	11.6	2.7	2.7	4.3	4.8	3.0	5.1	Eastleigh
12.2	12.6	11.0	11.4	2.8	2.9	4.8	5.4	3.3	5.4	Fareham
8.8	8.7	6.7	6.7	1.6	1.8	3.0	3.6	2.1	3.7	Gosport
10.1	10.1	8.9	9.1	2.1	2.1	2.9	3.0	1.8	3.0	Hart
11.9	12.7	11.5	12.0	3.1	3.4	5.7	6.5	3.9	6.1	Havant
16.2	17.6	17.3	18.3	4.7	5.1	8.8	10.1	7.6	12.0	New Forest
12.0	11.1	7.9	7.5	1.7	1.7	2.5	3.1	1.8	3.2	Rushmoor
12.8	13.0	11.4	11.8	2.8	2.7	4.3	4.4	2.7	5.1	Test Valley
11.2	11.7	11.1	11.2	2.6	2.8	4.5	4.8	3.3	5.8	Winchester
141.8	**146.3**	**130.4**	**134.4**	**33.7**	**35.0**	**55.2**	**62.8**	**40.3**	**70.2**	**Kent**
11.5	11.8	10.2	10.5	2.6	2.6	4.1	4.6	3.1	4.9	Ashford
12.5	13.0	12.0	13.2	3.3	3.6	5.7	6.7	4.9	8.7	Canterbury
10.4	10.4	7.7	7.7	2.0	2.0	3.2	3.6	2.1	3.5	Dartford
10.5	11.1	10.4	10.7	2.8	2.8	4.6	5.4	3.5	6.1	Dover
10.9	10.8	9.1	9.4	2.4	2.5	4.0	4.2	2.3	4.2	Gravesham
15.5	15.7	14.6	14.7	3.6	3.7	5.4	6.0	3.8	6.4	Maidstone
11.5	12.4	11.7	12.0	2.9	3.0	4.8	5.4	3.3	5.5	Sevenoaks
9.8	10.0	9.3	9.7	2.6	2.8	4.4	5.0	3.6	6.4	Shepway
13.7	13.6	12.4	12.3	3.1	3.1	4.8	5.2	3.1	5.3	Swale
11.7	12.6	11.8	12.5	3.3	3.6	5.9	7.2	5.3	9.2	Thanet
12.4	13.0	10.8	11.0	2.6	2.9	4.4	4.7	2.7	4.5	Tonbridge and Malling
11.6	12.0	10.3	10.6	2.6	2.5	4.0	4.6	2.8	5.5	Tunbridge Wells
71.3	**70.9**	**56.5**	**57.0**	**13.7**	**13.8**	**22.1**	**23.6**	**15.9**	**26.5**	**Oxfordshire**
16.9	16.6	12.6	12.4	2.9	2.9	4.5	4.9	3.1	5.2	Cherwell
14.3	13.7	9.4	9.8	2.4	2.4	4.2	4.4	3.3	5.6	Oxford
15.5	15.8	13.0	13.1	3.2	3.3	5.0	5.3	3.5	5.9	South Oxfordshire
13.2	13.4	11.9	11.8	2.9	2.8	4.6	4.9	3.2	5.2	Vale of White Horse
11.4	11.4	9.6	9.8	2.4	2.4	3.8	4.1	2.7	4.7	West Oxfordshire

Table 2.1 - *continued*

Area	All ages		Under 1		1 - 4		5 - 15		16 - 29	
	Males	Females	Males	Females	Males	Females	Males	Females	Males	Females
Surrey	**516.9**	**542.6**	**6.1**	**5.8**	**26.2**	**24.6**	**74.0**	**69.4**	**84.5**	**83.1**
Elmbridge	58.9	63.1	0.8	0.7	3.4	3.1	8.9	8.2	8.2	8.5
Epsom and Ewell	32.4	34.6	0.4	0.3	1.6	1.6	4.6	4.3	5.1	5.1
Guildford	64.1	65.7	0.7	0.7	3.0	2.7	8.4	7.9	13.8	12.6
Mole Valley	39.0	41.3	0.5	0.4	1.9	1.8	5.4	5.3	5.4	5.1
Reigate and Banstead	62.2	64.5	0.8	0.8	3.3	2.9	9.1	8.4	9.9	9.6
Runnymede	38.0	40.1	0.4	0.4	1.8	1.7	4.9	4.6	7.4	8.1
Spelthorne	44.4	46.0	0.5	0.5	2.1	2.1	6.1	5.8	7.1	7.0
Surrey Heath	39.7	40.6	0.5	0.4	2.1	1.9	6.0	5.6	6.1	5.9
Tandridge	38.2	41.1	0.5	0.5	1.9	1.9	5.7	5.6	5.3	5.5
Waverley	56.1	59.5	0.6	0.6	2.7	2.6	8.3	7.5	8.8	8.3
Woking	43.9	46.0	0.6	0.5	2.3	2.3	6.5	6.3	7.4	7.5
West Sussex	**361.1**	**393.2**	**4.1**	**3.8**	**17.5**	**16.9**	**52.6**	**49.4**	**54.2**	**54.3**
Adur	28.4	31.3	0.3	0.3	1.4	1.3	4.2	4.0	4.2	4.0
Arun	66.4	74.7	0.6	0.6	2.9	2.8	8.9	8.2	9.0	9.1
Chichester	50.3	56.2	0.5	0.5	2.2	2.1	7.0	6.6	7.4	7.3
Crawley	48.9	50.8	0.7	0.6	2.7	2.7	7.3	7.2	9.1	9.7
Horsham	59.4	62.9	0.7	0.6	3.0	2.9	9.3	8.7	8.2	8.2
Mid Sussex	61.8	65.6	0.7	0.7	3.1	3.0	9.5	8.6	9.2	9.0
Worthing	45.8	51.8	0.5	0.5	2.2	2.1	6.4	6.1	7.0	7.0
SOUTH WEST	**2,399.8**	**2,534.4**	**25.8**	**24.5**	**111.9**	**106.9**	**343.9**	**325.6**	**397.6**	**380.8**
Bath and North East Somerset UA	**82.2**	**86.9**	**0.9**	**0.8**	**3.8**	**3.5**	**11.1**	**10.7**	**15.8**	**15.3**
Bournemouth UA	**78.6**	**85.0**	**0.8**	**0.7**	**3.5**	**3.1**	**9.7**	**9.4**	**16.2**	**16.1**
Bristol, City of UA	**185.8**	**194.9**	**2.5**	**2.4**	**9.4**	**9.1**	**25.3**	**23.9**	**43.5**	**44.1**
North Somerset UA	**91.8**	**97.0**	**0.9**	**0.9**	**4.4**	**4.2**	**13.2**	**12.2**	**13.0**	**12.6**
Plymouth UA	**117.7**	**123.2**	**1.3**	**1.2**	**5.5**	**5.3**	**17.5**	**16.3**	**23.7**	**22.1**
Poole UA	**66.1**	**72.3**	**0.7**	**0.7**	**2.9**	**2.9**	**9.3**	**9.2**	**10.0**	**10.5**
South Gloucestershire UA	**121.6**	**124.4**	**1.4**	**1.4**	**6.5**	**6.0**	**18.5**	**17.4**	**19.9**	**19.3**
Swindon UA	**89.6**	**90.6**	**1.1**	**1.1**	**4.6**	**4.5**	**13.6**	**12.8**	**16.4**	**15.7**
Torbay UA	**61.9**	**68.0**	**0.6**	**0.5**	**2.7**	**2.5**	**8.9**	**8.3**	**9.0**	**8.8**
Cornwall and the Isles of Scilly	**242.9**	**259.1**	**2.4**	**2.2**	**11.1**	**10.4**	**34.3**	**32.4**	**35.6**	**34.3**
Caradon	38.6	41.1	0.4	0.3	1.6	1.6	5.5	5.3	5.5	5.2
Carrick	42.0	46.0	0.4	0.4	1.9	1.7	5.8	5.5	6.5	6.4
Kerrier	45.2	47.5	0.5	0.4	2.2	2.1	6.4	6.0	6.8	6.4
North Cornwall	39.2	41.5	0.4	0.4	1.8	1.7	5.6	5.3	5.4	5.2
Penwith	30.2	32.8	0.3	0.3	1.4	1.3	4.2	3.9	4.2	4.0
Restormel	46.7	49.1	0.5	0.4	2.2	2.1	6.7	6.3	7.0	6.9
Isles of Scilly	1.1	1.1	0.0	0.0	0.0	0.1	0.1	0.1	0.2	0.2
Devon	**340.7**	**364.9**	**3.3**	**3.1**	**14.5**	**13.7**	**47.8**	**45.3**	**53.8**	**49.5**
East Devon	59.3	66.4	0.5	0.5	2.4	2.2	7.6	7.5	8.3	7.6
Exeter	54.0	57.2	0.6	0.5	2.3	2.2	7.0	6.5	13.5	13.1
Mid Devon	34.2	35.7	0.4	0.4	1.6	1.6	5.2	4.8	4.9	4.6
North Devon	42.6	45.1	0.4	0.4	1.9	1.8	6.2	5.9	6.3	5.7
South Hams	39.7	42.3	0.4	0.3	1.6	1.5	5.8	5.4	5.4	4.6
Teignbridge	58.0	63.2	0.6	0.6	2.6	2.4	8.4	8.0	7.7	7.4
Torridge	28.9	30.2	0.3	0.3	1.2	1.1	4.1	4.0	4.2	3.8
West Devon	24.0	24.9	0.2	0.2	0.9	0.9	3.4	3.2	3.4	2.8
Dorset	**189.1**	**202.4**	**1.7**	**1.6**	**7.8**	**7.4**	**26.5**	**24.9**	**26.2**	**23.6**
Christchurch	21.2	23.8	0.2	0.2	0.8	0.7	2.7	2.5	2.6	2.5
East Dorset	40.1	43.9	0.3	0.3	1.6	1.6	5.4	5.1	4.8	4.5
North Dorset	30.9	31.1	0.3	0.2	1.4	1.2	4.7	4.4	5.3	4.0
Purbeck	21.5	22.9	0.2	0.2	0.9	0.9	3.2	2.9	2.8	2.7
West Dorset	44.1	48.3	0.4	0.4	1.9	1.8	6.1	5.9	5.6	5.4
Weymouth and Portland	31.3	32.4	0.3	0.3	1.3	1.3	4.3	4.2	5.1	4.6
Gloucestershire	**275.8**	**289.2**	**3.2**	**3.1**	**13.3**	**12.7**	**40.7**	**38.3**	**43.8**	**43.4**
Cheltenham	53.4	56.6	0.6	0.6	2.4	2.4	7.2	7.0	10.7	10.8
Cotswold	39.1	41.3	0.4	0.4	1.9	1.7	5.5	5.1	5.4	5.0
Forest of Dean	39.1	41.0	0.4	0.4	1.8	1.9	5.8	5.3	5.7	5.9
Gloucester	54.0	55.9	0.7	0.7	3.0	2.8	8.6	8.3	9.1	9.2
Stroud	53.0	55.0	0.6	0.6	2.5	2.4	8.2	7.4	7.6	7.1
Tewkesbury	37.2	39.3	0.4	0.4	1.8	1.7	5.4	5.1	5.3	5.4

United Kingdom by countries and, within England, Government Office Regions; counties/unitary authorities, districts and London boroughs in Great Britain/district council areas in Northern Ireland

thousands

30-44		45-59		60-64		65-74		75 and over		Area
Males	Females	Males	Females	Males	Females	Males	Females	Males	Females	
121.6	**125.8**	**107.2**	**107.4**	**25.0**	**26.3**	**41.4**	**46.7**	**30.9**	**53.4**	**Surrey**
14.3	15.5	12.4	12.3	2.7	3.0	4.6	5.1	3.7	6.5	Elmbridge
7.4	7.8	7.0	7.1	1.6	1.7	2.6	3.0	2.1	3.6	Epsom and Ewell
14.7	14.7	12.4	12.6	2.9	3.1	4.8	5.5	3.5	6.0	Guildford
8.5	8.9	8.6	8.8	2.1	2.3	3.7	4.1	2.8	4.7	Mole Valley
15.1	15.1	12.8	12.6	2.9	3.0	4.7	5.5	3.8	6.6	Reigate and Banstead
9.0	9.0	7.4	7.2	1.8	1.8	3.1	3.4	2.2	3.9	Runnymede
10.9	11.1	8.8	8.5	2.3	2.5	3.9	4.3	2.6	4.2	Spelthorne
9.8	10.1	8.5	8.5	2.0	2.0	3.0	3.2	1.7	3.0	Surrey Heath
8.8	9.3	8.5	8.6	1.9	2.0	3.2	3.7	2.5	4.0	Tandridge
12.3	12.9	12.0	12.6	2.9	3.0	4.7	5.3	3.7	6.8	Waverley
10.9	11.3	8.7	8.5	1.9	2.0	3.1	3.7	2.3	4.0	Woking
79.0	**82.1**	**72.9**	**76.0**	**18.5**	**20.1**	**33.7**	**40.3**	**28.7**	**50.3**	**West Sussex**
5.9	6.2	5.8	6.1	1.5	1.8	2.8	3.5	2.4	4.2	Adur
13.0	13.7	13.2	14.0	4.0	4.4	7.6	9.3	7.1	12.6	Arun
9.6	10.5	10.6	11.4	3.0	3.3	5.5	6.6	4.6	7.8	Chichester
12.3	12.0	8.5	8.6	1.7	1.9	3.7	4.3	2.9	3.8	Crawley
13.8	14.4	12.8	13.0	2.9	3.1	4.9	5.5	3.8	6.3	Horsham
14.0	14.6	13.5	14.0	3.1	3.2	5.0	5.8	3.6	6.8	Mid Sussex
10.4	10.6	8.6	8.9	2.2	2.5	4.2	5.3	4.3	8.8	Worthing
518.9	**532.0**	**485.3**	**499.5**	**128.2**	**132.5**	**218.4**	**245.6**	**169.9**	**286.9**	**SOUTH WEST**
17.7	**18.1**	**16.3**	**16.8**	**4.0**	**4.3**	**7.0**	**8.1**	**5.8**	**9.4**	**Bath and North East Somerset UA**
16.9	**16.7**	**14.1**	**14.4**	**3.7**	**4.0**	**6.8**	**8.5**	**6.8**	**12.1**	**Bournemouth UA**
42.5	**43.0**	**31.6**	**31.3**	**7.8**	**7.8**	**12.9**	**14.9**	**10.3**	**18.4**	**Bristol, City of UA**
19.7	**20.2**	**20.2**	**20.4**	**5.1**	**5.4**	**8.4**	**9.5**	**6.9**	**11.7**	**North Somerset UA**
26.6	**26.9**	**21.9**	**22.7**	**5.5**	**6.1**	**9.2**	**10.8**	**6.6**	**12.0**	**Plymouth UA**
14.6	**15.1**	**13.3**	**13.8**	**3.6**	**3.7**	**6.3**	**7.5**	**5.4**	**8.9**	**Poole UA**
29.8	**29.9**	**24.2**	**24.3**	**6.1**	**6.1**	**9.5**	**10.4**	**5.9**	**9.7**	**South Gloucestershire UA**
23.1	**22.3**	**16.2**	**16.0**	**4.0**	**4.0**	**6.4**	**7.2**	**4.3**	**7.0**	**Swindon UA**
12.3	**12.9**	**12.8**	**13.7**	**3.7**	**4.0**	**6.5**	**7.3**	**5.6**	**10.0**	**Torbay UA**
48.1	**50.6**	**53.4**	**56.1**	**15.0**	**15.2**	**24.9**	**27.0**	**18.3**	**30.9**	**Cornwall and Isles of Scilly**
7.8	8.3	8.9	9.4	2.3	2.3	3.8	4.1	2.8	4.7	Caradon
8.0	8.8	9.0	9.7	2.5	2.7	4.4	4.8	3.5	6.1	Carrick
9.3	9.7	9.7	10.1	2.8	2.8	4.5	4.8	3.1	5.3	Kerrier
7.6	7.9	8.5	9.0	2.6	2.5	4.2	4.5	3.0	5.0	North Cornwall
5.7	6.1	7.0	7.4	1.9	2.0	3.2	3.6	2.4	4.3	Penwith
9.4	9.6	10.1	10.3	2.8	2.9	4.7	5.1	3.3	5.5	Restormel
0.2	0.2	0.3	0.2	0.1	0.1	0.1	0.1	0.1	0.1	Isles of Scilly
67.2	**71.5**	**72.0**	**75.4**	**19.9**	**20.8**	**34.7**	**39.0**	**27.6**	**46.5**	**Devon**
10.3	11.4	12.2	13.4	3.8	4.3	7.3	8.6	6.8	11.0	East Devon
11.7	11.6	9.5	9.8	2.5	2.5	3.9	4.9	3.0	6.0	Exeter
7.1	7.5	7.4	7.5	2.0	2.0	3.3	3.5	2.4	3.9	Mid Devon
8.3	8.9	9.3	9.6	2.5	2.7	4.5	4.7	3.2	5.4	North Devon
7.7	8.6	9.0	9.6	2.4	2.5	4.1	4.5	3.2	5.2	South Hams
11.8	12.8	12.4	13.1	3.3	3.6	6.1	6.9	5.1	8.5	Teignbridge
5.5	5.8	6.5	6.7	1.9	1.8	3.1	3.3	2.2	3.5	Torridge
4.8	4.9	5.6	5.8	1.5	1.5	2.4	2.6	1.7	3.0	West Devon
36.9	**38.7**	**39.5**	**41.9**	**11.1**	**12.1**	**21.4**	**24.1**	**17.9**	**28.2**	**Dorset**
3.9	4.2	4.1	4.5	1.3	1.6	2.8	3.4	2.8	4.3	Christchurch
7.3	8.2	8.8	9.5	2.5	2.8	5.1	5.5	4.3	6.3	East Dorset
6.2	6.2	6.0	6.3	1.6	1.7	2.9	3.4	2.4	3.7	North Dorset
4.3	4.6	4.7	4.9	1.3	1.4	2.4	2.7	1.8	2.8	Purbeck
8.4	9.0	9.4	10.0	2.7	3.0	5.3	5.9	4.3	7.1	West Dorset
6.8	6.5	6.6	6.6	1.8	1.7	2.9	3.3	2.3	4.0	Weymouth and Portland
62.1	**63.6**	**56.8**	**56.9**	**14.4**	**14.3**	**23.4**	**26.2**	**17.9**	**30.7**	**Gloucestershire**
12.5	12.1	9.8	9.9	2.5	2.5	4.2	4.9	3.6	6.5	Cheltenham
8.3	8.9	8.7	8.7	2.1	2.2	3.8	4.2	3.0	5.0	Cotswold
8.3	8.5	8.7	8.8	2.3	2.2	3.5	3.8	2.4	4.2	Forest of Dean
13.2	13.5	9.9	9.7	2.4	2.5	4.1	4.5	3.0	4.9	Gloucester
11.5	12.0	11.7	11.7	2.9	2.8	4.5	5.0	3.5	6.0	Stroud
8.4	8.6	8.0	8.1	2.1	2.1	3.3	3.7	2.5	4.1	Tewkesbury

Table 2.1 - *continued*

Area	All ages		Under 1		1 - 4		5 - 15		16 - 29	
	Males	Females	Males	Females	Males	Females	Males	Females	Males	Females
Somerset	**242.3**	**256.4**	**2.4**	**2.4**	**11.2**	**11.1**	**36.0**	**33.9**	**36.6**	**34.7**
Mendip	50.8	53.2	0.5	0.5	2.4	2.4	8.0	7.5	7.6	7.2
Sedgemoor	51.7	54.3	0.6	0.5	2.4	2.4	7.8	7.2	7.4	7.2
South Somerset	73.8	77.2	0.8	0.8	3.5	3.4	10.7	10.4	11.3	10.3
Taunton Deane	49.3	53.2	0.5	0.5	2.2	2.2	7.4	6.8	8.1	7.9
West Somerset	16.6	18.4	0.1	0.1	0.6	0.6	2.1	2.0	2.2	2.1
Wiltshire	**213.7**	**219.8**	**2.6**	**2.5**	**10.9**	**10.5**	**31.6**	**30.6**	**34.3**	**31.0**
Kennet	37.6	37.3	0.5	0.4	1.9	1.8	5.4	5.3	6.8	5.3
North Wiltshire	62.1	63.3	0.8	0.7	3.3	3.2	9.6	9.1	9.5	8.9
Salisbury	56.1	58.5	0.7	0.7	2.7	2.6	8.0	7.8	9.2	8.3
West Wiltshire	57.9	60.6	0.7	0.7	2.9	3.0	8.6	8.4	8.8	8.5
WALES	**1,404.1**	**1,499.1**	**16.3**	**15.3**	**69.4**	**66.1**	**215.4**	**204.2**	**238.7**	**242.3**
Unitary Authorities										
Blaenau Gwent	34.0	36.0	0.4	0.4	1.6	1.6	5.6	5.4	5.6	5.5
Bridgend	62.5	66.2	0.7	0.8	3.1	3.0	9.6	9.1	10.2	10.0
Caerphilly	82.6	87.0	1.1	1.0	4.5	4.1	13.3	12.6	14.1	14.4
Cardiff	145.7	159.5	1.8	1.8	7.8	7.5	22.7	21.4	32.4	36.3
Carmarthenshire	83.6	90.1	1.0	0.9	3.9	3.7	12.5	11.9	12.6	13.1
Ceredigion	36.8	38.5	0.4	0.4	1.5	1.3	4.8	4.5	7.4	7.8
Conwy	52.3	57.5	0.5	0.5	2.3	2.3	7.5	7.1	7.5	7.2
Denbighshire	44.6	48.5	0.5	0.4	2.2	2.0	6.7	6.5	6.8	6.7
Flintshire	72.9	75.7	0.8	0.8	3.7	3.5	11.1	10.5	12.4	11.7
Gwynedd	56.0	60.8	0.7	0.7	2.8	2.6	8.1	7.7	9.8	10.2
Isle of Anglesey	32.3	34.4	0.4	0.3	1.5	1.4	4.8	4.6	4.9	4.9
Merthyr Tydfil	26.9	29.0	0.3	0.3	1.3	1.3	4.5	4.3	4.5	4.6
Monmouthshire	41.5	43.5	0.4	0.4	1.9	1.9	6.5	6.0	5.7	5.4
Neath Port Talbot	64.9	69.5	0.7	0.7	2.9	2.9	9.9	9.3	10.3	10.3
Newport	65.8	71.2	0.8	0.8	3.7	3.6	11.1	10.7	10.5	11.2
Pembrokeshire	54.5	58.5	0.7	0.6	2.8	2.5	8.3	8.2	7.7	8.1
Powys	62.5	63.9	0.6	0.6	2.9	2.6	9.2	8.6	8.7	7.8
Rhondda, Cynon, Taff	112.4	119.5	1.3	1.2	5.7	5.5	17.7	16.7	20.1	20.4
Swansea	108.1	115.1	1.2	1.2	4.9	4.7	15.6	14.6	20.1	20.0
Torfaen	44.0	46.9	0.5	0.4	2.2	2.1	7.3	6.9	7.1	7.1
The Vale of Glamorgan	57.4	61.9	0.7	0.6	3.0	3.0	9.4	8.9	8.8	9.0
Wrexham	62.8	65.8	0.8	0.7	3.1	2.9	9.3	8.6	11.3	10.9
SCOTLAND	**2,433.7**	**2,630.5**	**26.3**	**25.7**	**115.4**	**108.8**	**355.5**	**338.6**	**438.5**	**444.0**
Aberdeen City	103.7	108.2	1.1	1.0	4.3	3.9	12.5	12.0	24.1	23.8
Aberdeenshire	112.5	114.5	1.2	1.1	5.6	5.1	17.7	16.9	17.1	16.1
Angus	52.4	55.9	0.5	0.5	2.4	2.2	7.5	7.2	8.4	7.8
Argyll & Bute	44.9	46.4	0.4	0.4	2.0	1.8	6.4	6.1	6.9	6.1
Clackmannanshire	23.2	24.8	0.2	0.3	1.1	1.1	3.7	3.6	3.9	3.8
Dumfries & Galloway	71.3	76.5	0.7	0.6	3.1	3.0	10.5	9.9	10.0	9.9
Dundee City	69.0	76.4	0.7	0.7	3.0	2.9	9.4	9.0	14.9	15.2
East Ayrshire	57.9	62.4	0.6	0.6	2.7	2.5	9.0	8.5	9.6	9.6
East Dunbartonshire	52.0	56.3	0.5	0.5	2.4	2.3	8.4	7.8	8.4	8.2
East Lothian	43.0	47.2	0.5	0.5	2.3	2.2	6.8	6.5	5.8	6.1
East Renfrewshire	42.6	46.8	0.5	0.4	2.3	2.1	7.1	6.7	6.5	6.3
Edinburgh, City of	215.0	234.1	2.2	2.1	9.1	8.6	26.2	24.9	50.2	52.6
Eilean Siar	13.1	13.4	0.1	0.1	0.5	0.5	1.9	1.8	1.9	1.7
Falkirk	70.0	75.2	0.8	0.8	3.5	3.2	10.2	9.8	12.2	12.1
Fife	167.8	182.0	1.9	1.8	7.9	7.5	25.1	24.2	29.0	29.4
Glasgow City	273.0	305.7	3.2	3.1	13.1	12.3	38.4	36.3	58.9	64.5
Highland	102.3	106.6	1.1	1.0	4.9	4.3	15.2	14.3	15.9	14.5
Inverclyde	40.1	44.1	0.4	0.4	1.9	1.7	6.0	5.8	6.9	6.9
Midlothian	38.7	42.3	0.5	0.5	2.0	2.0	6.2	5.8	5.9	6.1
Moray	43.5	43.5	0.5	0.4	2.1	1.9	6.4	6.1	7.4	6.3

United Kingdom by countries and, within England, Government Office Regions; counties/unitary authorities, districts and London boroughs in Great Britain/district council areas in Northern Ireland

thousands

30-44		45-59		60-64		65-74		75 and over		Area
Males	Females	Males	Females	Males	Females	Males	Females	Males	Females	
51.3	**52.4**	**50.4**	**52.5**	**13.5**	**13.7**	**23.2**	**25.9**	**17.8**	**30.0**	**Somerset**
11.2	11.5	10.8	11.0	2.7	2.7	4.3	4.8	3.3	5.6	Mendip
11.0	11.3	10.9	11.3	3.0	3.0	5.0	5.5	3.7	6.1	Sedgemoor
15.7	15.5	15.2	15.7	4.0	4.2	7.1	7.9	5.5	9.1	South Somerset
10.5	11.1	9.9	10.5	2.5	2.6	4.6	5.3	3.5	6.2	Taunton Deane
2.9	3.1	3.6	3.9	1.2	1.2	2.2	2.4	1.8	2.9	West Somerset
50.2	**50.0**	**42.7**	**43.5**	**10.8**	**11.1**	**17.8**	**19.3**	**12.9**	**21.4**	**Wiltshire**
8.8	8.5	7.3	7.5	1.9	1.8	3.0	3.2	2.1	3.6	Kennet
15.3	15.5	12.6	12.4	2.9	3.0	4.7	5.1	3.3	5.4	North Wiltshire
12.8	12.8	11.0	11.4	2.8	3.1	5.1	5.6	3.7	6.3	Salisbury
13.2	13.3	11.7	12.2	3.1	3.1	5.1	5.4	3.8	6.2	West Wiltshire
294.9	**311.1**	**282.1**	**288.6**	**75.3**	**77.9**	**123.7**	**140.5**	**88.3**	**153.0**	**WALES**
										Unitary Authorities
7.2	7.7	6.8	6.7	1.9	1.8	2.9	3.3	2.0	3.7	Blaenau Gwent
14.2	14.5	12.5	12.8	3.3	3.6	5.3	6.1	3.6	6.3	Bridgend
18.1	18.7	16.3	16.8	4.3	4.3	6.6	7.5	4.3	7.5	Caerphilly
31.8	34.4	24.9	25.5	6.0	6.4	10.3	12.4	8.0	13.9	Cardiff
16.4	17.8	18.0	18.3	4.9	5.0	8.3	9.1	6.0	10.4	Carmarthenshire
6.8	6.7	7.8	7.8	2.1	2.3	3.5	3.7	2.7	4.1	Ceredigion
10.2	11.0	10.5	11.0	3.1	3.6	5.9	6.8	4.6	8.1	Conwy
8.9	9.5	9.2	9.5	2.6	2.7	4.3	5.0	3.4	6.2	Denbighshire
16.5	17.0	14.9	15.4	4.0	4.0	5.7	6.3	3.8	6.4	Flintshire
10.8	11.3	11.5	11.6	3.2	3.5	5.4	6.3	3.7	6.9	Gwynedd
6.3	6.5	7.0	7.3	2.0	2.1	3.2	3.6	2.1	3.7	Isle of Anglesey
5.8	6.3	5.3	5.4	1.5	1.5	2.3	2.6	1.4	2.7	Merthyr Tydfil
8.9	9.2	9.3	9.4	2.4	2.4	3.9	4.3	2.7	4.5	Monmouthshire
14.0	14.7	13.3	13.6	3.6	3.7	6.0	6.8	4.2	7.6	Neath Port Talbot
14.5	15.7	12.6	12.9	3.3	3.4	5.4	6.3	3.8	6.6	Newport
10.5	11.1	11.5	12.2	3.5	3.4	5.7	6.1	3.7	6.2	Pembrokeshire
12.5	12.8	13.6	13.8	3.7	3.6	6.3	6.7	4.9	7.3	Powys
24.0	25.1	22.2	22.4	5.8	5.9	9.3	10.7	6.3	11.5	Rhondda, Cynon, Taff
22.4	23.3	20.8	21.9	5.8	6.0	10.0	11.2	7.4	12.3	Swansea
9.4	10.0	9.0	9.0	2.2	2.3	3.6	4.6	2.6	4.4	Torfaen
12.0	13.3	12.0	12.4	3.0	3.1	4.8	5.5	3.7	6.1	The Vale of Glamorgan
13.6	14.3	13.0	13.0	3.2	3.3	5.0	5.7	3.5	6.4	Wrexham
563.4	**599.9**	**483.1**	**496.1**	**124.6**	**136.9**	**200.5**	**246.1**	**126.2**	**234.3**	**SCOTLAND**
24.6	24.2	19.5	18.9	4.7	5.0	7.9	9.8	5.1	9.7	Aberdeen City
26.5	27.1	24.9	24.0	5.6	5.6	8.5	9.4	5.4	9.2	Aberdeenshire
11.3	12.0	11.3	11.6	2.9	3.1	4.9	5.7	3.2	5.8	Angus
10.1	9.5	9.5	9.7	2.7	2.9	4.2	4.9	2.7	5.0	Argyll & Bute
5.4	5.6	4.9	5.0	1.2	1.3	1.8	2.1	1.0	2.1	Clackmannanshire
15.2	16.3	15.4	15.7	4.4	4.7	7.4	8.5	4.7	7.8	Dumfries & Galloway
14.4	16.1	12.7	13.0	3.7	4.0	6.1	7.8	4.2	7.7	Dundee City
13.2	14.2	11.8	12.1	3.1	3.4	4.9	5.9	3.0	5.5	East Ayrshire
11.4	12.8	10.9	11.7	2.8	3.2	4.5	5.3	2.5	4.5	East Dunbartonshire
10.1	11.0	8.8	9.1	2.2	2.5	3.8	4.6	2.6	4.7	East Lothian
9.7	11.0	8.8	9.4	2.2	2.4	3.4	4.4	2.2	4.1	East Renfrewshire
52.4	54.0	38.3	39.4	9.3	10.5	16.0	20.3	11.3	21.7	Edinburgh, City of
2.7	2.6	2.9	2.7	0.8	0.8	1.3	1.4	0.9	1.7	Eilean Siar
16.5	17.5	14.0	14.5	3.7	4.1	5.7	6.9	3.5	6.4	Falkirk
38.1	40.3	34.1	35.5	8.6	9.3	14.0	16.8	9.2	17.0	Fife
65.8	71.4	46.8	47.2	12.7	14.5	21.1	29.1	13.1	27.4	Glasgow City
22.3	23.8	22.5	22.4	5.7	6.1	9.1	10.4	5.6	9.7	Highland
9.1	10.0	8.1	8.3	2.1	2.4	3.4	4.4	2.1	4.1	Inverclyde
8.9	10.0	8.1	8.5	2.0	2.2	3.2	3.8	1.9	3.3	Midlothian
10.6	9.7	8.4	8.4	2.2	2.3	3.7	4.3	2.3	4.0	Moray

Population in administrative areas (Series VS no 28, PP1 no. 24)

Table 2.1 - *continued*

Area	All ages		Under 1		1 - 4		5 - 15		16 - 29	
	Males	Females	Males	Females	Males	Females	Males	Females	Males	Females
North Ayrshire	64.2	71.6	0.7	0.7	3.0	2.9	10.2	9.7	10.5	11.0
North Lanarkshire	154.0	167.2	1.8	1.8	8.0	7.6	23.6	22.5	28.7	29.1
Orkney Islands	9.5	9.7	0.1	0.1	0.4	0.4	1.4	1.4	1.3	1.3
Perth & Kinross	65.2	69.8	0.7	0.7	3.0	2.8	9.3	9.1	9.9	9.1
Renfrewshire	82.5	90.3	0.9	0.9	4.0	3.7	12.2	11.6	13.9	14.2
Scottish Borders	51.5	55.5	0.5	0.5	2.3	2.2	7.5	7.2	6.9	7.1
Shetland Islands	11.1	10.9	0.1	0.1	0.6	0.5	1.8	1.7	1.9	1.7
South Ayrshire	53.4	58.7	0.5	0.5	2.3	2.2	7.5	7.3	8.3	8.2
South Lanarkshire	144.3	158.1	1.6	1.5	7.0	6.8	22.0	20.6	25.0	25.4
Stirling	41.2	45.0	0.4	0.4	2.0	1.9	6.0	5.7	7.4	8.0
West Dunbartonshire	44.2	49.2	0.5	0.5	2.2	2.1	6.9	6.5	7.8	8.1
West Lothian	76.8	82.2	1.0	0.9	4.3	4.3	12.5	11.9	13.0	13.7
NORTHERN IRELAND	**824.4**	**864.9**	**11.1**	**10.4**	**47.9**	**45.3**	**144.8**	**137.6**	**164.6**	**163.4**
Antrim	24.6	24.2	0.4	0.3	1.5	1.4	4.1	3.9	5.1	4.5
Ards	35.8	37.6	0.4	0.4	2.0	1.8	5.6	5.3	6.3	6.2
Armagh	27.0	27.4	0.4	0.4	1.6	1.6	4.9	4.7	5.5	5.3
Ballymena	28.7	30.1	0.4	0.4	1.5	1.5	4.7	4.4	5.3	5.2
Ballymoney	13.3	13.7	0.2	0.2	0.8	0.8	2.3	2.2	2.5	2.5
Banbridge	20.8	20.7	0.3	0.3	1.3	1.1	3.6	3.2	3.9	3.8
Belfast	129.8	147.4	1.6	1.6	6.8	6.5	22.1	21.4	29.6	32.2
Carrickfergus	18.3	19.4	0.3	0.2	1.0	0.9	3.1	3.0	3.3	3.4
Castlereagh	31.7	34.8	0.4	0.4	1.8	1.7	5.1	5.0	4.9	5.0
Coleraine	27.0	29.4	0.4	0.3	1.6	1.5	4.5	4.3	5.0	5.8
Cookstown	16.3	16.4	0.2	0.2	1.0	0.9	3.1	3.0	3.5	3.3
Craigavon	39.9	41.0	0.6	0.5	2.4	2.3	7.5	6.7	7.4	7.3
Derry	51.4	54.0	0.8	0.7	3.4	3.2	10.2	9.9	11.4	11.4
Down	31.8	32.3	0.4	0.4	1.9	1.7	5.8	5.6	6.3	5.7
Dungannon	23.7	24.2	0.3	0.3	1.5	1.4	4.4	4.3	4.9	4.7
Fermanagh	28.9	28.8	0.4	0.3	1.6	1.6	5.1	4.8	5.6	5.3
Larne	15.1	15.7	0.2	0.2	0.8	0.7	2.5	2.3	2.5	2.6
Limavady	16.7	15.9	0.2	0.2	1.0	1.0	3.0	2.8	3.8	3.3
Lisburn	53.2	55.8	0.7	0.7	3.3	3.0	9.7	9.1	10.2	10.0
Magherafelt	20.1	19.8	0.3	0.3	1.3	1.2	3.7	3.6	4.4	4.0
Moyle	7.8	8.1	0.1	0.1	0.5	0.4	1.4	1.3	1.4	1.4
Newry and Mourne	43.3	44.1	0.7	0.6	2.8	2.7	8.4	7.9	8.7	8.5
Newtownabbey	38.8	41.4	0.5	0.4	2.1	2.0	6.3	6.0	7.3	7.6
North Down	37.1	39.5	0.4	0.4	1.8	1.7	5.6	5.3	6.9	6.2
Omagh	24.2	23.9	0.3	0.3	1.5	1.4	4.5	4.3	5.3	4.7
Strabane	19.2	19.1	0.3	0.3	1.3	1.2	3.5	3.4	3.9	3.7

United Kingdom by countries and, within England, Government Office Regions; counties/unitary authorities, districts and London boroughs in Great Britain/district council areas in Northern Ireland

thousands

30-44		45-59		60-64		65-74		75 and over		Area
Males	Females	Males	Females	Males	Females	Males	Females	Males	Females	
14.1	16.0	13.3	14.2	3.6	3.9	5.5	6.8	3.3	6.3	North Ayrshire
36.7	39.7	29.4	31.2	7.7	8.7	11.8	14.6	6.3	11.9	North Lanarkshire
2.2	2.1	2.1	2.1	0.6	0.6	0.8	0.9	0.6	0.9	Orkney Islands
14.3	15.1	13.9	14.4	3.6	4.0	6.2	7.3	4.2	7.4	Perth & Kinross
19.7	21.6	16.6	17.5	4.3	4.9	6.8	8.4	3.9	7.4	Renfrewshire
11.7	12.2	11.1	11.3	2.9	3.3	5.0	5.7	3.5	6.0	Scottish Borders
2.5	2.4	2.5	2.2	0.5	0.5	0.8	0.9	0.5	1.0	Shetland Islands
11.5	12.5	11.5	12.0	3.1	3.4	5.2	6.3	3.6	6.4	South Ayrshire
33.7	36.9	29.0	30.7	7.5	8.5	11.7	14.9	6.8	12.7	South Lanarkshire
9.1	9.9	8.4	8.7	2.2	2.4	3.3	4.1	2.1	3.9	Stirling
10.2	11.3	8.8	9.1	2.1	2.5	3.5	4.7	2.1	4.4	West Dunbartonshire
19.6	21.0	15.1	15.7	3.7	3.9	5.0	5.9	2.6	4.8	West Lothian
184.5	**191.5**	**144.1**	**146.4**	**35.5**	**38.2**	**55.6**	**67.8**	**36.2**	**64.4**	**NORTHERN IRELAND**
6.0	5.7	4.3	4.3	1.0	1.1	1.4	1.5	0.8	1.4	Antrim
8.2	8.6	7.3	7.4	1.7	1.8	2.5	3.0	1.8	3.0	Ards
6.0	5.8	4.7	4.6	1.1	1.2	1.7	2.1	1.1	1.8	Armagh
6.4	6.6	5.5	5.6	1.4	1.5	2.1	2.6	1.4	2.4	Ballymena
3.1	3.0	2.3	2.3	0.6	0.6	0.9	1.1	0.6	1.0	Ballymoney
5.0	4.9	3.7	3.5	0.8	0.9	1.3	1.5	0.9	1.5	Banbridge
27.0	30.8	20.9	22.2	5.7	6.4	9.6	13.1	6.5	13.2	Belfast
4.5	4.7	3.4	3.4	0.8	0.9	1.3	1.6	0.8	1.4	Carrickfergus
7.9	8.7	5.6	5.9	1.5	1.7	2.7	3.4	1.8	3.1	Castlereagh
6.0	6.3	5.0	5.0	1.3	1.5	2.0	2.4	1.3	2.3	Coleraine
3.5	3.5	2.7	2.8	0.6	0.7	1.0	1.1	0.7	1.1	Cookstown
9.3	9.3	6.8	7.0	1.7	1.9	2.6	3.1	1.6	2.8	Craigavon
11.3	12.3	8.1	8.4	2.0	2.0	2.8	3.3	1.4	2.8	Derry
7.0	7.1	5.6	5.6	1.3	1.4	2.0	2.4	1.4	2.4	Down
5.1	5.0	4.0	3.9	1.0	1.0	1.5	1.9	1.0	1.6	Dungannon
6.2	6.0	5.3	5.1	1.3	1.2	2.0	2.2	1.5	2.3	Fermanagh
3.6	3.6	3.0	2.9	0.7	0.8	1.2	1.4	0.7	1.3	Larne
3.9	3.5	2.7	2.6	0.7	0.6	0.8	1.0	0.5	0.9	Limavady
12.5	13.3	9.3	9.6	2.2	2.4	3.3	4.0	2.0	3.6	Lisburn
4.4	4.3	3.2	3.2	0.7	0.8	1.2	1.4	0.8	1.2	Magherafelt
1.6	1.7	1.5	1.4	0.4	0.4	0.6	0.7	0.4	0.7	Moyle
9.5	9.6	7.1	7.1	1.7	1.8	2.6	3.2	1.6	2.7	Newry and Mourne
8.9	9.5	7.1	7.3	1.8	2.0	3.0	3.6	1.7	3.1	Newtownabbey
8.1	8.5	7.7	8.0	1.7	1.8	2.8	3.4	2.2	4.1	North Down
5.2	5.2	4.1	3.9	0.9	1.0	1.4	1.6	0.9	1.6	Omagh
4.2	4.1	3.2	3.1	0.8	0.8	1.2	1.4	0.7	1.1	Strabane

Table 2.2 Mid-2001 population estimates: estimated resident population by broad age-group and sex

Area	All ages		Under 1		1 - 4		5-15		16-29	
	Males	Females	Males	Females	Males	Females	Males	Females	Males	Females
UNITED KINGDOM	**28,611.3**	**30,225.3**	**338.4**	**324.0**	**1,442.6**	**1,372.3**	**4,292.4**	**4,085.0**	**5,142.5**	**5,166.7**
ENGLAND AND WALES	**25,354.7**	**26,729.8**	**300.9**	**287.9**	**1,279.2**	**1,218.2**	**3,792.1**	**3,608.8**	**4,540.1**	**4,559.3**
ENGLAND	**23,950.6**	**25,230.8**	**284.6**	**272.6**	**1,209.8**	**1,152.1**	**3,576.7**	**3,404.6**	**4,301.4**	**4,317.0**
Northern and Yorkshire	**3,017.3**	**3,202.4**	**33.9**	**32.3**	**146.9**	**140.7**	**458.2**	**437.9**	**532.9**	**539.3**
Bradford	225.4	242.5	3.3	3.2	13.5	13.3	38.5	37.5	43.0	45.7
Calderdale and Kirklees	282.1	299.3	3.5	3.4	15.6	15.0	44.7	42.7	48.7	50.2
County Durham and Darlington	287.0	304.6	2.9	2.8	13.2	12.8	42.1	40.1	48.3	48.9
East Riding and Hull	272.6	285.6	3.0	2.7	12.9	12.4	41.4	39.3	46.2	45.9
Gateshead and South Tyneside	166.5	177.4	1.9	1.8	8.0	7.5	25.0	23.7	27.5	27.8
Leeds	346.0	369.5	4.0	3.8	16.8	16.0	52.2	49.8	73.3	75.4
Newcastle and North Tyneside	217.5	234.1	2.5	2.4	10.1	9.6	31.2	29.6	43.2	45.0
North Cumbria	153.4	160.0	1.5	1.5	6.8	6.5	22.1	20.8	23.6	23.2
Northumberland	150.0	157.4	1.4	1.4	6.6	6.3	21.7	20.4	22.7	21.8
North Yorkshire	365.2	386.2	3.8	3.5	16.3	15.6	52.0	50.1	60.8	57.0
Sunderland	136.6	144.2	1.5	1.3	6.5	6.0	20.9	19.8	25.9	25.8
Tees	261.7	279.6	3.0	2.9	13.1	12.4	42.8	41.1	44.8	46.5
Wakefield	153.4	162.0	1.7	1.7	7.5	7.1	23.6	23.0	24.9	26.1
Trent	**2,492.4**	**2,597.2**	**27.7**	**26.5**	**120.7**	**114.4**	**373.2**	**354.1**	**435.1**	**428.8**
Barnsley	106.2	112.0	1.2	1.1	5.2	4.9	16.2	15.6	17.0	17.1
Doncaster	140.1	146.8	1.6	1.6	6.9	6.7	21.9	21.1	23.2	22.6
Leicestershire	454.1	470.6	5.5	5.2	22.9	21.5	68.2	65.0	85.4	84.0
Lincolnshire	317.1	330.6	3.1	3.0	14.4	13.5	46.4	43.7	48.4	46.9
North Derbyshire	180.2	187.4	1.8	1.7	8.3	7.7	26.4	24.4	26.8	26.0
North Nottinghamshire	192.7	201.4	2.1	1.9	9.2	8.9	29.4	28.0	30.0	29.9
Nottingham	306.1	315.6	3.4	3.2	14.3	13.5	44.4	42.0	63.4	60.8
Rotherham	120.8	127.6	1.5	1.4	6.1	6.0	19.1	18.1	19.4	20.0
Sheffield	250.6	262.5	2.8	2.7	12.2	11.4	35.2	33.6	52.1	51.0
Southern Derbyshire	273.1	283.4	3.2	3.0	13.7	13.0	41.5	39.4	46.2	46.3
South Humber	151.5	159.4	1.7	1.7	7.5	7.3	24.4	23.1	23.4	24.1
Eastern	**2,642.2**	**2,752.7**	**31.1**	**30.0**	**133.5**	**126.6**	**389.9**	**372.6**	**447.0**	**439.8**
Bedfordshire	281.5	284.9	3.7	3.7	15.3	14.7	44.9	42.0	50.3	50.8
Cambridgeshire	350.4	359.6	4.2	4.0	17.4	16.7	51.0	48.3	66.8	64.8
Hertfordshire	505.6	529.3	6.4	6.3	26.7	25.6	76.0	73.9	84.3	85.5
Norfolk	388.5	409.4	3.9	3.7	17.1	16.5	52.7	50.7	61.2	60.7
North Essex	447.3	465.8	5.2	4.9	22.8	20.8	65.2	62.4	74.4	70.6
South Essex	340.5	362.7	4.0	4.0	18.1	16.9	51.9	49.3	56.9	57.6
Suffolk	328.4	341.0	3.6	3.3	16.1	15.3	48.3	46.0	53.0	49.8
London	**3,479.5**	**3,708.5**	**49.6**	**47.4**	**194.1**	**186.9**	**496.7**	**475.8**	**757.5**	**800.8**
Barking and Havering	186.5	202.6	2.3	2.2	10.3	10.1	29.9	28.1	32.6	34.3
Barnet, Enfield and Haringey	385.4	421.0	5.5	5.3	21.3	20.9	57.6	55.8	81.0	86.5
Bexley, Bromley and Greenwich	350.8	379.4	4.8	4.6	19.5	18.6	53.3	50.9	62.0	64.5
Brent and Harrow	228.8	243.0	2.9	2.8	11.5	11.2	33.5	31.9	51.1	51.2
Camden and Islington	180.1	194.4	2.5	2.5	9.2	8.8	21.5	21.1	45.4	52.6
Croydon	159.7	171.8	2.2	2.1	9.3	9.0	25.9	24.1	29.3	31.0
Ealing, Hammersmith & Hounslow	331.8	347.9	4.7	4.4	17.6	16.9	44.3	42.8	78.0	81.4
East London and The City	319.9	331.6	5.6	5.3	21.4	20.7	52.9	51.0	78.9	86.3
Hillingdon	117.6	125.5	1.6	1.5	6.8	6.2	18.1	17.3	22.1	23.7
Kensington & Chelsea and Westminster	165.1	175.7	2.3	2.2	7.6	7.4	15.4	14.7	39.0	43.1
Kingston and Richmond	155.8	164.6	2.2	2.1	8.5	8.2	20.1	19.3	30.6	30.4
Lambeth, Southwark and Lewisham	372.1	389.5	5.7	5.4	21.3	20.7	50.7	49.5	90.1	93.2
Merton, Sutton and Wandsworth	303.0	326.4	4.5	4.2	16.6	15.8	39.2	36.9	70.2	75.9
Redbridge and Waltham Forest	222.9	235.1	3.0	2.9	13.1	12.3	34.3	32.4	47.3	46.6

Note: Figures are displayed for health authority boundaries as at 1 April 2001.

United Kingdom by countries and, within England, Health Regional Office areas, and health authorities

thousands

30 - 44		45-59		60-64		65-74		75 and over		Area
Males	Females	Males	Females	Males	Females	Males	Females	Males	Females	
6,544.1	**6,747.2**	**5,520.0**	**5,624.2**	**1,409.3**	**1,470.4**	**2,303.8**	**2,636.0**	**1,618.2**	**2,799.6**	**UNITED KINGDOM**
5,797.0	**5,955.7**	**4,892.6**	**4,981.8**	**1,249.2**	**1,295.3**	**2,047.7**	**2,322.1**	**1,455.8**	**2,500.9**	**ENGLAND AND WALES**
5,502.1	**5,644.6**	**4,610.6**	**4,693.1**	**1,173.9**	**1,217.3**	**1,924.0**	**2,181.6**	**1,367.4**	**2,347.8**	**ENGLAND**
671.9	**697.9**	**595.8**	**600.2**	**154.1**	**161.8**	**253.0**	**293.9**	**170.5**	**298.5**	**Northern and Yorkshire**
49.1	51.2	40.4	40.7	10.2	10.6	16.1	20.0	11.3	20.4	Bradford
64.3	66.3	56.3	56.4	13.5	13.7	20.6	24.6	14.9	27.0	Calderdale and Kirklees
64.8	66.8	59.0	59.5	15.6	16.2	24.9	28.8	16.1	28.6	County Durham and Darlington
59.6	60.2	54.9	55.1	14.4	14.9	24.1	27.5	16.0	27.7	East Riding and Hull
37.6	39.6	32.3	32.2	9.1	9.5	15.4	18.0	9.8	17.3	Gateshead and South Tyneside
76.6	80.7	61.6	62.6	15.9	17.1	26.6	31.1	19.0	32.9	Leeds
48.6	50.8	40.9	40.6	10.1	11.3	18.2	21.8	12.8	22.9	Newcastle and North Tyneside
34.4	34.5	32.5	32.4	8.7	9.0	14.4	16.1	9.3	16.1	North Cumbria
32.7	33.9	33.4	33.7	8.4	8.9	13.9	15.6	9.4	15.3	Northumberland
79.9	83.4	75.9	77.7	20.0	21.0	32.7	37.1	23.8	40.8	North Yorkshire
30.4	31.4	26.1	26.9	6.9	7.4	11.7	13.7	6.7	11.8	Sunderland
57.6	62.3	51.3	51.6	13.5	14.1	22.2	25.4	13.4	23.5	Tees
36.2	36.8	31.4	30.8	7.7	8.1	12.2	14.2	8.2	14.2	Wakefield
559.3	**569.7**	**493.2**	**493.8**	**126.8**	**130.9**	**209.4**	**234.7**	**146.9**	**244.3**	**Trent**
24.7	25.5	21.6	21.4	5.5	5.8	8.8	10.3	6.0	10.2	Barnsley
31.7	32.4	27.5	27.9	7.0	7.7	12.2	13.9	7.9	13.0	Doncaster
102.2	104.2	88.6	88.9	21.6	22.1	35.0	39.2	24.8	40.5	Leicestershire
66.7	68.9	65.9	67.4	18.2	18.9	31.7	33.9	22.2	34.4	Lincolnshire
40.3	41.2	39.0	38.6	9.9	10.0	16.1	18.2	11.7	19.5	North Derbyshire
43.9	44.9	40.0	40.0	10.3	10.4	16.5	18.4	11.4	18.9	North Nottinghamshire
69.4	70.0	56.2	55.9	14.0	14.6	24.0	27.0	17.1	28.6	Nottingham
27.7	28.5	24.4	24.4	6.4	6.6	9.8	11.3	6.5	11.2	Rotherham
56.5	55.9	45.0	45.5	12.3	12.8	19.8	23.1	14.6	26.4	Sheffield
62.5	63.7	54.2	53.4	13.6	13.5	22.3	24.7	15.9	26.4	Southern Derbyshire
33.7	34.5	30.7	30.4	8.1	8.3	13.1	14.8	8.8	15.2	South Humber
603.0	**607.4**	**524.7**	**533.7**	**131.6**	**135.0**	**221.4**	**244.6**	**159.9**	**262.9**	**Eastern**
67.6	66.3	53.1	52.3	13.1	12.6	20.5	21.6	13.0	20.9	Bedfordshire
81.0	81.3	68.4	68.9	16.5	16.4	26.3	28.8	18.7	30.3	Cambridgeshire
123.8	125.4	98.8	98.6	22.9	23.9	39.1	44.0	27.5	46.1	Hertfordshire
81.2	81.4	80.1	82.8	22.6	23.2	39.6	43.4	30.1	47.1	Norfolk
100.6	101.9	91.1	93.3	22.5	23.5	37.8	41.8	27.6	46.5	North Essex
77.7	79.5	67.5	70.2	16.6	17.7	27.9	32.3	19.9	35.0	South Essex
71.0	71.5	65.8	67.6	17.5	17.7	30.1	32.7	23.0	37.0	Suffolk
909.5	**939.7**	**561.1**	**594.9**	**136.8**	**145.4**	**219.0**	**249.1**	**155.1**	**268.7**	**London**
43.0	45.3	34.1	35.7	8.2	9.1	14.8	18.4	11.2	19.3	Barking and Havering
97.7	105.1	63.4	69.2	15.4	17.1	25.6	28.8	17.8	32.2	Barnet, Enfield and Haringey
85.7	89.9	64.1	67.5	15.6	17.1	26.2	31.4	19.7	35.0	Bexley, Bromley and Greenwich
55.7	60.0	38.1	40.9	9.9	10.7	15.9	17.2	10.1	17.1	Brent and Harrow
51.7	52.1	26.7	28.4	6.3	6.7	10.0	11.2	6.8	11.1	Camden and Islington
39.5	43.8	28.7	30.1	6.6	7.2	10.5	12.0	7.7	12.5	Croydon
88.6	90.1	52.5	55.1	13.2	13.4	19.3	21.3	13.4	22.4	Ealing, Hammersmith & Hounslow
82.8	82.0	41.2	42.3	10.6	10.6	16.6	16.9	10.1	16.5	East London and The City
28.9	30.2	21.0	21.4	5.0	5.5	8.4	9.6	5.8	10.1	Hillingdon
47.7	46.6	28.0	30.5	7.0	7.4	10.4	11.9	7.7	11.9	Kensington & Chelsea and Westminster
41.5	42.0	29.3	30.6	6.1	6.2	9.5	11.1	8.0	14.6	Kingston and Richmond
109.0	109.4	50.4	53.4	12.3	13.2	19.3	22.0	13.4	22.7	Lambeth, Southwark and Lewisham
83.2	85.3	46.5	50.2	11.4	12.0	18.1	21.4	13.3	24.7	Merton, Sutton and Wandsworth
54.6	57.8	37.0	39.5	9.1	9.2	14.3	15.9	10.1	18.6	Redbridge and Waltham Forest

Table 2.2 - *continued*

Area	All ages		Under 1		1 - 4		5-15		16-29	
	Males	Females	Males	Females	Males	Females	Males	Females	Males	Females
South East	**4,220.8**	**4,416.6**	**49.5**	**47.4**	**212.0**	**201.4**	**626.4**	**591.1**	**732.1**	**716.5**
Berkshire	399.2	401.0	5.1	4.8	20.9	20.0	59.9	55.6	78.2	74.4
Buckinghamshire	338.0	348.7	4.5	4.3	18.5	17.4	52.9	50.0	57.7	57.6
East Kent	283.4	308.8	3.1	2.9	14.0	13.2	43.1	41.3	46.0	47.7
East Surrey	204.8	216.5	2.5	2.4	10.7	10.1	30.1	28.4	30.0	29.7
East Sussex, Brighton and Hove	353.6	387.6	3.8	3.7	16.3	15.7	48.4	46.3	57.9	60.5
Isle of Wight, Portsmouth and South East Hampshire	328.7	346.1	3.7	3.5	15.5	14.6	48.0	45.8	57.6	54.5
Northamptonshire	311.2	319.3	3.8	3.6	16.0	15.3	49.0	45.7	53.2	51.7
North and Mid Hampshire	275.5	279.3	3.4	3.1	14.3	13.6	41.1	38.6	49.4	45.1
Oxfordshire	299.5	306.4	3.5	3.4	14.7	14.2	42.7	39.9	59.1	57.1
Southampton and South West Hampshire	269.2	279.5	2.8	2.8	12.5	11.7	38.2	36.5	53.7	50.7
West Kent	484.5	504.2	5.8	5.6	25.7	24.1	76.6	72.6	80.6	79.7
West Surrey	312.1	326.1	3.6	3.5	15.5	14.6	43.9	41.0	54.4	53.4
West Sussex	361.1	393.2	4.1	3.8	17.5	16.9	52.6	49.4	54.2	54.3
South West	**2,399.8**	**2,534.4**	**25.8**	**24.5**	**111.9**	**106.9**	**343.9**	**325.6**	**397.6**	**380.8**
Avon	481.4	503.3	5.7	5.5	24.0	22.7	68.1	64.2	92.1	91.2
Cornwall and Isles of Scilly	242.9	259.1	2.4	2.2	11.1	10.4	34.3	32.4	35.6	34.3
Dorset	333.7	359.7	3.2	3.0	14.1	13.5	45.5	43.5	52.4	50.2
Gloucestershire	275.8	289.2	3.2	3.1	13.3	12.7	40.7	38.3	43.8	43.4
North and East Devon	235.6	252.3	2.3	2.2	10.0	9.6	32.4	30.9	39.3	36.7
Somerset	242.3	256.4	2.4	2.4	11.2	11.1	36.0	33.9	36.6	34.7
South and West Devon	284.7	303.9	2.8	2.6	12.6	11.9	41.7	38.9	47.1	43.7
Wiltshire	303.3	310.4	3.7	3.6	15.5	15.0	45.1	43.4	50.7	46.7
West Midlands	**2,575.5**	**2,691.6**	**30.4**	**29.4**	**132.8**	**125.4**	**397.5**	**379.6**	**453.3**	**450.5**
Birmingham	473.0	503.4	6.8	6.6	29.0	27.3	80.8	78.2	98.1	102.5
Coventry	149.1	151.7	1.8	1.8	7.7	7.4	23.3	21.7	32.7	30.5
Dudley	149.7	155.4	1.7	1.6	7.3	6.8	22.4	20.9	24.0	24.2
Herefordshire	85.4	89.5	0.8	0.8	4.0	3.8	12.6	11.9	11.9	11.6
North Staffordshire	222.9	234.2	2.5	2.4	10.4	9.8	32.1	30.8	40.3	40.6
Sandwell	136.4	146.3	1.7	1.8	7.4	7.2	22.0	21.3	23.8	25.2
Shropshire	218.2	223.6	2.4	2.3	10.7	10.3	32.9	30.9	37.0	33.3
Solihull	96.7	102.8	1.0	1.0	4.8	4.3	15.7	15.0	14.2	14.2
South Staffordshire	290.7	299.9	3.2	3.0	13.8	13.5	43.4	42.0	47.3	44.7
Walsall	123.2	130.2	1.6	1.6	6.7	6.5	19.5	19.1	20.8	21.2
Warwickshire	248.4	257.8	2.7	2.5	12.0	11.3	35.9	34.0	40.0	39.6
Wolverhampton	115.8	120.6	1.4	1.3	6.0	5.5	18.2	17.1	21.4	21.6
Worcestershire	266.0	276.2	2.8	2.7	13.0	11.7	38.8	36.6	41.8	41.3
North West	**3,123.1**	**3,327.4**	**36.5**	**35.1**	**158.0**	**149.8**	**490.8**	**467.9**	**545.8**	**560.5**
Bury and Rochdale	187.7	198.2	2.3	2.2	10.2	10.0	31.3	29.6	31.7	32.4
East Lancashire	252.5	265.0	3.3	3.1	14.3	13.7	43.6	41.0	42.6	43.9
Liverpool	209.8	229.4	2.5	2.3	10.2	9.7	32.7	30.9	44.5	49.3
Manchester	191.8	201.1	2.5	2.5	10.0	9.6	29.7	28.3	50.9	52.0
Morecambe Bay	148.8	159.6	1.6	1.5	6.8	6.2	21.3	20.1	25.3	26.3
North Cheshire	151.1	158.4	1.8	1.7	7.7	7.6	23.8	22.8	25.8	25.9
North-West Lancashire	217.3	233.7	2.3	2.3	10.3	9.5	31.5	30.6	35.2	36.4
St Helens and Knowsley	156.8	170.5	1.9	1.8	8.2	7.5	26.1	25.5	26.6	27.5
Salford and Trafford	208.3	217.7	2.4	2.3	10.3	9.5	31.9	30.1	37.9	37.4
Sefton	133.4	149.5	1.3	1.4	6.2	5.9	21.5	20.8	20.6	20.7
South Cheshire	328.3	346.0	3.6	3.5	15.8	15.0	48.9	46.0	50.2	51.0
South Lancashire	153.0	159.9	1.7	1.5	7.3	6.8	23.2	22.4	25.0	24.6
Stockport	137.3	147.3	1.5	1.5	6.8	6.6	21.1	20.2	21.0	21.9
West Pennine	224.5	238.5	2.8	2.8	12.5	11.8	37.4	35.2	37.4	39.5
Wigan and Bolton	275.2	287.6	3.3	3.2	14.3	13.5	42.8	41.4	48.4	47.6
Wirral	147.1	165.1	1.7	1.7	7.2	6.9	24.0	23.1	22.7	24.1

Note: Figures are displayed for health authority boundaries as at 1 April 2001.

United Kingdom by countries and, within England, Health Regional Office areas, and health authorities

thousands

30 - 44		45-59		60-64		65-74		75 and over		Area
Males	Females	Males	Females	Males	Females	Males	Females	Males	Females	
970.7	**987.4**	**838.4**	**849.2**	**204.6**	**210.6**	**335.9**	**380.3**	**251.2**	**432.9**	**South East**
99.4	97.5	75.1	73.8	17.1	16.8	26.3	28.7	17.3	29.3	Berkshire
80.0	82.4	69.4	69.6	15.6	15.5	23.7	25.6	15.7	26.3	Buckinghamshire
58.8	61.6	56.4	59.3	15.3	16.0	25.6	30.2	21.2	36.6	East Kent
47.4	49.6	44.1	44.3	10.0	10.7	16.7	19.1	13.3	22.4	East Surrey
78.8	79.8	68.4	71.0	18.3	19.5	32.6	39.2	29.1	51.8	East Sussex, Brighton and Hove
72.7	74.2	64.1	65.5	16.7	17.5	28.6	33.3	21.7	37.2	Isle of Wight, Portsmouth and South East Hampshire
73.1	74.5	63.0	63.1	14.6	14.4	22.4	24.7	16.1	26.3	Northamptonshire
66.3	66.1	55.7	55.8	12.9	12.9	19.5	21.2	12.9	22.9	North and Mid Hampshire
71.3	70.9	56.5	57.0	13.7	13.8	22.1	23.6	15.9	26.5	Oxfordshire
57.4	58.3	52.0	51.8	12.9	13.2	22.1	25.1	17.6	29.5	Southampton and South West Hampshire
112.2	114.3	97.6	98.9	24.0	24.6	37.8	41.7	24.2	42.7	West Kent
74.2	76.2	63.1	63.1	15.0	15.7	24.7	27.6	17.6	31.0	West Surrey
79.0	82.1	72.9	76.0	18.5	20.1	33.7	40.3	28.7	50.3	West Sussex
518.9	**532.0**	**485.3**	**499.5**	**128.2**	**132.5**	**218.4**	**245.6**	**169.9**	**286.9**	**South West**
109.6	111.2	92.2	92.8	23.0	23.6	37.7	42.8	28.9	49.2	Avon
48.1	50.6	53.4	56.1	15.0	15.2	24.9	27.0	18.3	30.9	Cornwall and Isles of Scilly
68.5	70.5	67.0	70.0	18.5	19.8	34.6	40.1	30.1	49.2	Dorset
62.1	63.6	56.8	56.9	14.4	14.3	23.4	26.2	17.9	30.7	Gloucestershire
46.4	48.9	48.8	51.0	13.8	14.4	23.8	26.9	18.8	31.8	North and East Devon
51.3	52.4	50.4	52.5	13.5	13.7	23.2	25.9	17.8	30.0	Somerset
59.7	62.4	57.9	60.9	15.4	16.5	26.6	30.2	20.9	36.7	South and West Devon
73.2	72.3	58.9	59.5	14.7	15.0	24.2	26.5	17.2	28.4	Wiltshire
570.3	**582.0**	**502.7**	**504.4**	**132.1**	**134.7**	**211.8**	**238.2**	**144.5**	**247.4**	**West Midlands**
101.7	106.6	76.6	78.2	20.6	21.4	34.5	39.3	24.9	43.2	Birmingham
32.5	31.8	25.0	25.0	6.6	7.2	11.1	12.5	8.4	14.0	Coventry
33.7	33.9	30.6	30.1	8.4	8.5	13.3	15.0	8.2	14.5	Dudley
18.3	18.5	18.4	18.5	4.9	5.1	8.3	9.3	6.1	9.9	Herefordshire
49.0	49.6	45.4	44.5	11.5	11.8	18.8	22.1	13.0	22.5	North Staffordshire
30.7	31.9	24.5	24.4	7.2	7.0	11.3	13.6	7.8	14.0	Sandwell
48.5	48.7	44.8	45.4	11.7	11.9	18.2	19.8	12.2	20.8	Shropshire
21.3	22.4	20.4	21.2	5.1	5.2	8.4	9.9	5.7	9.6	Solihull
66.2	67.5	62.3	62.5	16.1	15.8	23.6	25.9	14.8	25.1	South Staffordshire
27.1	27.5	23.3	23.5	6.8	7.0	10.9	12.4	6.5	11.4	Walsall
56.8	57.3	52.7	52.8	13.2	13.4	20.9	22.7	14.3	24.2	Warwickshire
25.3	25.9	20.6	20.6	5.8	5.9	10.1	11.4	7.1	11.4	Wolverhampton
59.3	60.5	58.0	57.8	14.3	14.3	22.4	24.6	15.7	26.7	Worcestershire
698.4	**728.7**	**609.4**	**617.6**	**159.7**	**166.5**	**255.1**	**295.2**	**169.3**	**306.2**	**North West**
42.9	44.3	37.1	37.6	9.1	9.2	14.0	16.4	9.2	16.6	Bury and Rochdale
55.6	57.3	49.5	48.7	12.1	12.1	18.7	21.8	12.9	23.4	East Lancashire
45.7	49.4	36.2	38.0	10.1	10.6	17.5	20.1	10.3	19.1	Liverpool
41.3	42.3	28.5	27.9	7.5	8.0	12.5	14.4	8.9	16.1	Manchester
31.6	32.3	30.1	30.7	8.4	8.6	13.7	15.9	10.0	17.9	Morecambe Bay
35.6	36.9	30.4	31.0	7.6	7.7	11.4	12.6	7.0	12.2	North Cheshire
47.0	47.9	42.9	42.8	12.2	13.0	20.9	24.4	15.0	26.7	North-West Lancashire
35.2	38.8	29.9	31.2	8.3	8.8	13.5	15.8	7.2	13.5	St Helens and Knowsley
48.6	48.4	38.5	38.9	10.0	10.7	17.0	19.4	11.7	21.1	Salford and Trafford
28.6	31.8	26.4	28.2	7.3	8.7	13.1	15.8	8.5	16.2	Sefton
74.5	77.0	69.7	70.0	17.9	18.6	28.4	32.1	19.3	32.8	South Cheshire
34.6	35.5	32.7	33.3	8.5	8.5	12.4	13.8	7.7	13.6	South Lancashire
32.3	33.5	28.1	28.2	7.3	7.5	11.5	13.6	7.8	14.2	Stockport
51.8	53.9	43.8	44.1	11.3	11.5	16.3	19.0	11.2	20.7	West Pennine
62.6	64.2	55.3	55.1	14.3	14.4	20.8	23.7	13.4	24.6	Wigan and Bolton
30.6	35.2	30.4	31.8	7.8	8.6	13.5	16.4	9.3	17.5	Wirral

Table 2.2 - *continued*

Area	All ages		Under 1		1 - 4		5-15		16-29	
	Males	Females	Males	Females	Males	Females	Males	Females	Males	Females
WALES	**1,404.1**	**1,499.1**	**16.3**	**15.3**	**69.4**	**66.1**	**215.4**	**204.2**	**238.7**	**242.3**
North Wales	320.8	342.7	3.7	3.5	15.6	14.8	47.6	45.0	52.7	51.6
Dyfed Powys	237.3	251.0	2.6	2.4	11.1	10.1	34.9	33.2	36.5	36.7
Morgannwg	235.6	250.8	2.6	2.6	11.0	10.6	35.0	33.0	40.7	40.2
Bro Taf	342.5	369.9	4.2	3.9	17.9	17.3	54.3	51.4	65.7	70.2
Gwent	267.9	284.6	3.2	2.9	13.8	13.3	43.7	41.6	43.1	43.6
SCOTLAND	**2,433.7**	**2,630.5**	**26.3**	**25.7**	**115.4**	**108.8**	**355.5**	**338.6**	**438.5**	**444.0**
Argyll & Clyde	202.1	218.6	2.1	2.1	9.6	8.9	30.1	28.7	33.7	33.2
Ayrshire & Arran	175.5	192.7	1.8	1.8	8.0	7.6	26.6	25.5	28.4	28.8
Borders	51.5	55.5	0.5	0.5	2.3	2.2	7.5	7.2	6.9	7.1
Dumfries & Galloway	71.3	76.5	0.7	0.6	3.1	3.0	10.5	9.9	10.0	9.9
Fife	167.7	182.0	1.9	1.8	7.9	7.5	25.1	24.2	28.9	29.4
Forth Valley	134.3	144.9	1.4	1.5	6.7	6.2	20.0	19.1	23.5	24.0
Grampian	259.7	266.2	2.7	2.5	12.0	10.9	36.6	35.0	48.5	46.2
Greater Glasgow	410.3	457.9	4.7	4.5	20.0	18.7	60.2	56.9	81.6	87.3
Highland	102.3	106.6	1.1	1.0	4.9	4.3	15.2	14.3	15.9	14.5
Lanarkshire	265.5	287.8	3.0	3.0	13.4	12.9	40.7	38.4	47.9	48.2
Lothian	373.4	405.6	4.1	4.0	17.7	17.2	51.7	49.1	74.9	78.6
Orkney	9.5	9.7	0.1	0.1	0.4	0.4	1.4	1.4	1.3	1.3
Shetland	11.1	10.9	0.1	0.1	0.6	0.5	1.8	1.7	1.9	1.7
Tayside	186.6	202.1	1.9	1.9	8.5	7.9	26.2	25.2	33.2	32.1
Western Isles	13.1	13.4	0.1	0.1	0.5	0.5	1.9	1.8	1.9	1.7
NORTHERN IRELAND	**824.4**	**864.9**	**11.1**	**10.4**	**47.9**	**45.3**	**144.8**	**137.6**	**164.6**	**163.4**
Eastern	319.5	347.4	4.0	3.9	17.5	16.4	53.9	51.7	64.2	65.4
Northern	209.9	218.3	2.9	2.5	12.0	11.4	35.6	33.9	40.2	40.1
Southern	154.7	157.5	2.3	2.0	9.6	9.1	28.9	26.8	30.4	29.6
Western	140.3	141.7	2.0	1.9	8.8	8.3	26.4	25.1	29.9	28.4

Note: Figures are displayed for health authority boundaries as at 1 April 2001.

United Kingdom by countries and, within England, Health Regional Office areas, and health authorities

thousands

30 - 44		45-59		60-64		65-74		75 and over		Area
Males	Females	Males	Females	Males	Females	Males	Females	Males	Females	
294.9	**311.1**	**282.1**	**288.6**	**75.3**	**77.9**	**123.7**	**140.5**	**88.3**	**153.0**	**WALES**
66.4	69.7	66.2	67.7	18.1	19.2	29.4	33.6	21.1	37.7	North Wales
46.1	48.4	50.9	52.1	14.1	14.3	23.8	25.6	17.3	28.2	Dyfed Powys
50.7	52.5	46.5	48.3	12.6	13.2	21.2	24.1	15.2	26.2	Morgannwg
73.6	79.2	64.4	65.8	16.4	16.8	26.7	31.2	19.4	34.3	Bro Taf
58.2	61.3	54.1	54.8	14.0	14.4	22.5	26.0	15.3	26.7	Gwent
563.4	**599.9**	**483.1**	**496.1**	**124.6**	**136.9**	**200.5**	**246.1**	**126.2**	**234.3**	**SCOTLAND**
47.1	50.0	41.3	42.7	10.9	12.2	17.1	21.2	10.3	19.5	Argyll & Clyde
38.8	42.7	36.6	38.3	9.8	10.8	15.6	19.1	9.9	18.2	Ayrshire & Arran
11.7	12.2	11.1	11.3	2.9	3.3	5.0	5.7	3.5	6.0	Borders
15.2	16.3	15.4	15.7	4.4	4.7	7.4	8.5	4.7	7.8	Dumfries & Galloway
38.1	40.3	34.1	35.5	8.6	9.3	14.0	16.8	9.1	17.0	Fife
30.9	32.9	27.3	28.2	7.1	7.7	10.8	13.0	6.7	12.3	Forth Valley
61.7	61.0	52.7	51.3	12.5	12.9	20.1	23.4	12.8	22.8	Grampian
96.4	106.2	74.8	77.3	19.7	22.6	32.7	43.5	20.1	40.7	Greater Glasgow
22.3	23.8	22.5	22.4	5.7	6.1	9.1	10.4	5.6	9.7	Highland
62.9	68.0	51.6	54.8	13.6	15.2	20.7	25.9	11.5	21.4	Lanarkshire
90.9	96.1	70.3	72.7	17.3	19.0	27.9	34.5	18.5	34.4	Lothian
2.2	2.1	2.1	2.1	0.6	0.6	0.8	0.9	0.6	0.9	Orkney
2.5	2.4	2.5	2.2	0.5	0.5	0.8	0.9	0.5	1.0	Shetland
40.0	43.3	37.9	39.0	10.2	11.1	17.2	20.8	11.6	20.8	Tayside
2.7	2.6	2.9	2.7	0.8	0.8	1.3	1.4	0.9	1.7	Western Isles
184.5	**191.5**	**144.1**	**146.4**	**35.5**	**38.2**	**55.6**	**67.8**	**36.2**	**64.4**	**NORTHERN IRELAND**
70.7	76.9	56.5	58.7	14.1	15.6	22.9	29.4	15.7	29.4	Eastern
48.0	48.8	38.0	38.2	9.4	10.3	14.6	17.1	9.3	15.9	Northern
35.0	34.6	26.3	26.2	6.4	6.8	9.8	11.8	6.1	10.5	Southern
30.9	31.2	23.4	23.2	5.6	5.6	8.2	9.5	5.1	8.6	Western

Table 3.1 Mid-year population estimates: components of population change mid-1981 to mid-2001

United Kingdom by countries and, within England, Government Office Regions; counties/unitary authorities; Northern Ireland council districts

thousands

Area	Resident population mid-1981	Change 1981-1991		Resident population mid-1991[1]	Change 1991-2001		Resident population mid-2001
		Natural change	Other changes[2]		Natural Change	Other changes[2]	
UNITED KINGDOM	**56,357.5**	**1,027.6**	**53.6**	**57,438.7**	**999.1**	**398.9**	**58,836.7**
ENGLAND AND WALES	**49,634.3**	**885.2**	**228.5**	**50,748.0**	**918.5**	**418.0**	**52,084.5**
ENGLAND	**46,820.8**	**860.2**	**194.0**	**47,875.0**	**917.9**	**388.4**	**49,181.3**
NORTH EAST	**2,636.2**	**22.2**	**-71.4**	**2,587.0**	**-0.3**	**-70.2**	**2,516.5**
Darlington UA	98.6	..	..	99.3	-0.2	-1.3	97.9
Hartlepool UA	94.9	..	..	91.1	1.4	-3.8	88.7
Middlesbrough UA	150.6	..	..	144.7	4.6	-14.6	134.8
Redcar and Cleveland UA	150.9	..	..	145.9	1.1	-7.9	139.2
Stockton-on-Tees UA	173.9	..	..	175.2	4.5	-1.1	178.6
Durham	512.8	..	..	501.5	-2.3	-5.5	493.7
Northumberland	299.3	-3.4	9.6	305.5	-4.8	6.7	307.4
Tyne and Wear (Met County)	1,155.2	3.1	-34.5	1,123.8	-4.7	-42.8	1,076.3
NORTH WEST	**6,940.3**	**83.6**	**-180.9**	**6,843.0**	**54.3**	**-165.8**	**6,731.5**
Blackburn with Darwen UA	142.6	..	..	137.4	7.1	-7.0	137.6
Blackpool UA	149.1	..	..	148.6	-5.6	-0.8	142.3
Halton UA	123.2	..	..	124.8	4.0	-10.6	118.2
Warrington UA	169.8	..	..	184.7	4.9	1.6	191.2
Cheshire	639.4	..	..	653.3	4.8	16.1	674.2
Cumbria	481.2	-4.1	9.2	486.3	-5.2	6.7	487.8
Greater Manchester (Met County)	2,619.1	49.9	-115.4	2,553.6	45.2	-115.9	2,482.8
Lancashire	1,093.8	..	..	1,116.4	0.1	19.3	1,135.8
Merseyside (Met County)	1,522.2	17.9	-102.1	1,438.0	-1.0	-75.4	1,361.7
YORKSHIRE AND THE HUMBER	**4,918.4**	**65.1**	**-47.4**	**4,936.1**	**66.9**	**-35.8**	**4,967.2**
East Riding of Yorkshire UA	271.6	..	..	294.4	-4.6	25.1	314.8
Kingston upon Hull, City of UA	273.7	..	..	263.3	6.9	-26.8	243.4
North East Lincolnshire UA	161.3	..	..	161.0	2.1	-5.2	158.0
North Lincolnshire UA	151.0	..	..	152.9	1.2	-1.1	153.0
York UA	165.4	..	..	172.3	0.7	8.3	181.3
North Yorkshire	511.5	..	..	541.8	-4.2	32.4	570.1
South Yorkshire (Met County)	1,317.1	13.6	-42.1	1,288.7	11.3	-33.4	1,266.5
West Yorkshire (Met County)	2,066.8	43.4	-48.5	2,061.7	53.5	-35.0	2,080.2
EAST MIDLANDS	**3,852.8**	**76.4**	**82.2**	**4,011.4**	**56.2**	**107.5**	**4,175.1**
Derby UA	217.4	..	..	222.9	6.3	-7.5	221.7
Leicester UA	283.1	..	..	281.5	12.3	-14.0	279.8
Nottingham UA	278.2	..	..	279.4	8.1	-20.5	267.0
Rutland UA	33.0	..	..	33.0	0.3	1.3	34.6
Derbyshire	696.9	..	..	713.5	0.9	20.4	734.9
Leicestershire	542.6	..	..	573.3	12.2	24.8	610.3
Lincolnshire	553.1	-1.2	36.7	588.6	-6.6	65.7	647.6
Northamptonshire	532.5	19.5	32.1	584.1	17.6	28.8	630.4
Nottinghamshire	716.1	..	..	735.1	5.1	8.6	748.8
WEST MIDLANDS	**5,186.6**	**138.3**	**-95.2**	**5,229.7**	**107.0**	**-69.6**	**5,267.1**
Herefordshire, County of UA	149.4	..	..	160.4	-1.1	15.5	174.9
Stoke-on-Trent UA	252.3	..	..	249.4	2.5	-11.4	240.4
Telford & Wrekin UA	125.5	..	..	141.3	7.7	9.5	158.5
Shropshire	255.3	..	..	268.7	-1.2	15.8	283.3
Staffordshire	766.9	..	..	791.6	8.6	6.9	807.1
Warwickshire	477.2	8.0	2.0	487.1	4.4	14.7	506.2
West Midlands (Met County)	2,673.1	86.1	-140.4	2,618.8	80.5	-145.0	2,554.4
Worcestershire	487.0	..	..	512.4	5.6	24.3	542.2

1 Mid-1991 population estimates for all areas (except Northern Ireland and its districts) have been revised in light of results of the 2001 Census.

2 This column is not an estimate of net civilian migration. It has been derived by subtraction using revised population estimates and natural change. Although the main component of these other changes is net civilian migration, this is not the only component. Changes to the non-civilian population and definitional differences are also included.

Table 3.1 - *continued*

United Kingdom by countries and, within England, Government Office Regions; counties/unitary authorities; Northern Ireland council districts

thousands

Area	Resident population mid-1981	Change 1981-1991		Resident population mid-1991[1]	Change 1991-2001		Resident population mid-2001
		Natural change	Other changes[2]		Natural change	Other changes[2]	
EAST	**4,854.1**	**127.5**	**139.5**	**5,121.1**	**113.0**	**160.8**	**5,394.9**
Luton UA	164.8	..	..	173.7	15.2	-4.6	184.3
Peterborough UA	133.8	..	..	154.2	8.3	-6.0	156.5
Southend-on-sea UA	157.6	..	..	161.2	-1.7	0.9	160.4
Thurrock UA	127.4	..	..	128.7	8.0	6.5	143.2
Bedfordshire	345.2	..	..	355.9	14.9	11.2	382.1
Cambridgeshire	455.6	..	..	510.6	13.7	29.3	553.6
Essex	1,198.0	..	..	1,249.1	20.6	43.0	1,312.7
Hertfordshire	967.2	31.5	-14.4	984.3	37.8	12.8	1,034.9
Norfolk	702.9	-3.9	55.3	754.3	-8.0	51.6	797.9
Suffolk	601.6	9.4	38.1	649.1	4.1	16.2	669.4
LONDON	**6,805.6**	**242.0**	**-218.3**	**6,829.3**	**407.0**	**-48.3**	**7,188.0**
SOUTH EAST	**7,245.4**	**111.2**	**272.6**	**7,629.2**	**129.7**	**248.0**	**8,006.9**
Bracknell Forest UA	84.7	..	..	98.1	7.3	4.3	109.6
Brighton & Hove UA	237.3	..	..	240.5	-2.8	10.4	248.1
Isle of Wight UA	118.1	..	..	125.9	-6.0	13.0	132.9
Medway UA	240.3	..	..	242.5	11.8	-4.6	249.7
Milton Keynes UA	126.0	..	..	178.3	15.0	14.3	207.6
Portsmouth UA	191.4	..	..	186.8	2.7	-2.6	186.9
Reading UA	137.4	..	..	134.8	8.3	0.2	143.2
Slough UA	97.6	..	..	105.4	9.3	4.4	119.1
Southampton UA	209.8	..	..	204.8	5.1	7.7	217.6
West Berkshire UA	122.6	..	..	138.8	6.4	-0.8	144.5
Windsor and Maidenhead UA	135.4	..	..	133.7	3.3	-3.4	133.5
Wokingham UA	117.0	..	..	141.1	8.3	0.9	150.4
Buckinghamshire	444.4	..	..	455.7	17.0	6.4	479.1
East Sussex	428.0	..	..	469.0	-20.8	44.9	493.1
Hampshire	1,088.2	..	..	1,178.7	26.2	35.9	1,240.8
Kent	1,244.2	..	..	1,285.8	9.1	36.2	1,331.1
Oxfordshire	541.8	22.4	11.9	576.1	22.5	7.3	605.9
Surrey	1,015.0	8.5	-0.2	1,023.3	19.8	16.3	1,059.5
West Sussex	666.3	-17.2	60.9	710.0	-12.9	57.2	754.3
SOUTH WEST	**4,381.4**	**-6.1**	**313.0**	**4,688.2**	**-15.9**	**261.8**	**4,934.2**
Bath and North East Somerset UA	161.5	..	..	163.1	-0.5	6.6	169.2
Bournemouth UA	143.4	..	..	157.3	-7.9	14.1	163.6
Bristol, City of UA	401.2	..	..	392.2	9.9	-21.3	380.8
North Somerset UA	162.9	..	..	179.2	-3.4	13.0	188.8
Plymouth UA	253.3	..	..	251.2	2.9	-13.2	241.0
Poole UA	120.3	..	..	134.0	-1.3	5.7	138.4
South Gloucestershire UA	203.1	..	..	222.2	11.9	12.0	246.0
Swindon UA	151.6	..	..	171.5	8.4	0.3	180.2
Torbay UA	113.1	..	..	122.0	-6.3	14.3	130.0
Cornwall and Isles of Scilly	426.4	-5.0	50.1	471.5	-10.2	40.8	502.1
Devon	599.2	..	..	658.5	-17.5	64.5	705.6
Dorset	334.9	..	..	365.2	-11.9	38.2	391.5
Gloucestershire	506.2	6.2	23.5	536.0	5.2	23.9	565.0
Somerset	430.7	-1.9	36.9	465.7	-4.1	37.1	498.7
Wiltshire	373.5	..	..	398.7	9.0	25.9	433.5

Table 3.1 - *continued*

United Kingdom by countries and, within England, Government Office Regions; counties/unitary authorities; Northern Ireland council districts

thousands

Area	Resident population mid-1981	Change 1981-1991		Resident population mid-1991[1]	Change 1991-2001		Resident population mid-2001
		Natural change	Other changes[2]		Natural Change	Other changes[2]	
WALES	**2,813.5**	**25.0**	**34.5**	**2,873.0**	**0.5**	**29.6**	**2,903.2**
Unitary Authorities							
Blaenau Gwent	75.7	..	..	72.7	-0.4	-2.2	70.0
Bridgend	126.2	..	..	129.5	0.9	-1.7	128.7
Caerphilly	171.8	..	..	170.6	3.3	-4.4	169.6
Cardiff	286.8	..	..	296.9	9.8	-1.5	305.2
Carmarthenshire	165.1	..	..	169.7	-4.9	8.9	173.7
Ceredigion	61.2	..	..	65.9	-1.5	10.9	75.3
Conwy	99.0	..	..	108.0	-5.3	7.1	109.8
Denbighshire	86.7	..	..	89.4	-2.8	6.5	93.1
Flintshire	138.6	..	..	142.0	2.8	3.7	148.6
Gwynedd	111.9	..	..	115.0	-1.1	2.9	116.8
Isle of Anglesey	68.0	..	..	69.1	-0.5	-1.9	66.7
Merthyr Tydfil	60.5	..	..	59.6	0.2	-3.8	56.0
Monmouthshire	76.5	..	..	80.2	-0.3	5.1	85.0
Neath Port Talbot	142.7	..	..	138.8	-2.5	-2.0	134.4
Newport	132.4	..	..	135.5	3.0	-1.4	137.0
Pembrokeshire	107.4	..	..	112.4	-0.5	1.0	113.0
Powys	112.2	..	..	119.7	-1.7	8.4	126.4
Rhondda, Cynon, Taff	238.4	..	..	234.9	0.7	-3.6	231.9
Swansea	229.3	..	..	229.7	-1.6	-5.0	223.2
Torfaen	90.7	..	..	91.0	1.4	-1.4	90.9
The Vale of Glamorgan	113.2	..	..	118.1	1.2	0.0	119.3
Wrexham	119.2	..	..	124.2	0.4	3.9	128.5
SCOTLAND	**5,180.2**	**27.2**	**-124.1**	**5,083.3**	**-6.3**	**-12.8**	**5,064.2**
Aberdeen City	212.5	..	..	214.1	1.0	-3.2	211.9
Aberdeenshire	188.9	..	..	215.9	5.7	5.4	226.9
Angus	105.6	..	..	108.4	-1.6	1.5	108.4
Argyll & Bute	90.9	..	..	93.5	-2.8	0.5	91.3
Clackmannanshire	48.2	..	..	48.1	0.7	-0.7	48.1
Dumfries & Galloway	145.5	..	..	147.2	-2.7	3.3	147.8
Dundee City	169.6	..	..	155.6	-2.2	-7.9	145.5
East Ayrshire	127.4	..	..	124.0	-0.6	-3.1	120.3
East Dunbartonshire	109.7	..	..	110.2	1.7	-3.6	108.3
East Lothian	80.7	..	..	84.4	-0.1	5.8	90.2
East Renfrewshire	80.2	..	..	85.8	1.9	1.8	89.4
Edinburgh, City of	446.0	..	..	436.3	-0.7	13.4	449.0
Eilean Siar	31.5	..	..	29.3	-1.4	-1.5	26.5
Falkirk	145.1	..	..	142.5	0.1	2.7	145.3
Fife	341.5	..	..	347.4	-0.4	2.8	349.8
Glasgow City	712.4	..	..	629.2	-9.3	-41.2	578.7
Highland	194.9	..	..	203.8	0.3	4.8	208.9
Inverclyde	101.2	..	..	91.4	-1.9	-5.4	84.2
Midlothian	83.6	..	..	79.5	1.3	0.1	81.0
Moray	83.5	..	..	84.0	1.2	1.8	87.0
North Ayrshire	137.3	..	..	138.1	-0.5	-1.8	135.8
North Lanarkshire	341.7	..	..	326.9	4.6	-10.4	321.2
Orkney Islands	19.2	..	..	19.5	-0.2	-0.1	19.2
Perth & Kinross	121.9	..	..	127.4	-2.7	10.3	135.0
Renfrewshire	185.1	..	..	175.7	0.2	-3.1	172.9
Scottish Borders	101.3	..	..	103.8	-2.3	5.4	107.0
Shetland Islands	26.3	..	..	22.5	0.5	-1.0	22.0
South Ayrshire	113.2	..	..	113.1	-3.5	2.5	112.2
South Lanarkshire	310.0	..	..	302.5	1.5	-1.7	302.3
Stirling	80.3	..	..	80.9	-0.2	5.5	86.2
West Dunbartonshire	105.8	..	..	97.3	-0.3	-3.7	93.3
West Lothian	139.2	..	..	145.0	6.1	7.9	159.0

1 Mid-1991 population estimates for all areas (except Northern Ireland and its districts) have been revised in light of results of the 2001 Census.
2 This column is not an estimate of net civilian migration. It has been derived by subtraction using revised population estimates and natural change. Although the main component of these other changes is net civilian migration, this is not the only component. Changes to the non-civilian population and definitional differences are also included.

Table 3.1 - *continued*

United Kingdom by countries and, within England, Government Office Regions; counties/unitary authorities; Northern Ireland council districts

thousands

Area	Resident population mid-1981	Change 1981-1991		Resident population mid-1991[1]	Change 1991-2001		Resident population mid-2001
		Natural change	Other changes[2]		Natural change	Other changes[2]	
NORTHERN IRELAND	**1,543.0**	**112.9**	**-48.6**	**1,607.3**	**87.0**	**-4.9**	**1,689.3**
Antrim	46.1	4.7	-5.1	45.6	3.8	-0.6	48.8
Ards	58.0	3.3	3.9	65.3	2.3	5.9	73.4
Armagh	49.3	3.6	-0.7	52.3	3.2	-1.0	54.5
Ballymena	55.0	3.5	-1.8	56.7	2.3	-0.2	58.8
Ballymoney	23.0	1.7	-0.4	24.2	1.2	1.5	27.0
Banbridge	30.1	1.6	1.9	33.6	2.2	5.7	41.5
Belfast	316.4	12.2	-35.6	292.9	6.6	-22.4	277.2
Carrickfergus	28.7	2.0	2.4	33.2	1.9	2.6	37.7
Castlereagh	60.9	0.3	0.3	61.5	2.6	2.5	66.5
Coleraine	46.9	2.5	3.2	52.6	2.3	1.5	56.4
Cookstown	28.4	2.7	0.1	31.2	2.3	-0.7	32.7
Craigavon	73.5	7.3	-5.4	75.4	5.3	0.2	80.9
Derry	90.2	13.0	-5.6	97.6	9.7	-2.0	105.3
Down	53.8	4.6	0.3	58.6	3.3	2.2	64.1
Dungannon	43.9	4.5	-2.9	45.5	3.1	-0.7	47.8
Fermanagh	52.1	3.6	-1.1	54.6	2.1	1.0	57.7
Larne	29.1	1.0	-0.5	29.6	1.0	0.2	30.8
Limavady	27.3	3.3	-1.0	29.6	2.8	0.2	32.6
Lisburn	85.2	8.1	7.7	101.0	8.0	0.0	109.0
Magherafelt	32.7	3.6	0.1	36.4	3.1	0.4	39.9
Moyle	14.5	0.8	-0.4	14.9	0.4	0.7	16.0
Newry and Mourne	77.5	9.0	-3.0	83.6	7.2	-3.4	87.4
Newtownabbey	72.5	5.5	-1.8	76.1	3.5	0.5	80.1
North Down	66.9	2.6	3.8	73.3	0.8	2.5	76.6
Omagh	44.8	4.4	-3.3	45.9	3.2	-1.1	48.1
Strabane	36.3	3.4	-3.6	36.2	2.8	-0.7	38.3

Table 3.2 Mid-year population estimates: components of population change, mid-1991 to mid-2001

United Kingdom by countries and, within England, Health Regional Office areas

thousands

Area	Mid-year population (thousands)			Change 1991-2001		Components of change 1991-2001	
	1991[1]	2000[1]	2001	Thousands	Percentage	Natural change	Other changes[2]
UNITED KINGDOM	**57,438.7**	**58,643.2**	**58,836.7**	**1,398.0**	**2.4**	**999.1**	**398.9**
ENGLAND AND WALES	**50,748.0**	**51,897.3**	**52,084.5**	**1,336.5**	**2.6**	**918.5**	**418.0**
ENGLAND	**47,875.0**	**48,997.3**	**49,181.3**	**1,306.3**	**2.7**	**917.9**	**388.4**
Northern and Yorkshire	6,235.1	6,209.0	6,219.7	-15.4	-0.2	49.1	-64.5
Trent	4,999.3	5,076.8	5,089.6	90.3	1.8	52.2	38.2
Eastern	5,121.1	5,374.9	5,394.9	273.8	5.3	113.0	160.8
London	6,829.3	7,104.4	7,188.0	358.7	5.3	407.0	-48.3
South East	8,213.3	8,607.3	8,637.3	424.1	5.2	147.3	276.7
South West	4,688.2	4,909.2	4,934.2	245.9	5.2	-15.9	261.8
West Midlands	5,229.7	5,260.0	5,267.1	37.4	0.7	107.0	-69.6
North West	6,559.1	6,455.8	6,450.5	-108.5	-1.7	58.2	-166.7
WALES	**2,873.0**	**2,900.1**	**2,903.2**	**30.2**	**1.0**	**0.5**	**29.6**
SCOTLAND	**5,083.3**	**5,062.9**	**5,064.2**	**-19.1**	**-0.4**	**-6.3**	**-12.8**
NORTHERN IRELAND	**1,607.3**	**1,682.9**	**1,689.3**	**82.0**	**5.1**	**87.0**	**-4.9**

Note: Figures are displayed for Health Regional Office area boundaries as at 1 April 2001.

1 Population estimates have been revised in light of results of the 2001 Census.
2 This column is not an estimate of net civilian migration. It has been derived by subtraction using revised population estimates and natural change. Although the main component of these other changes is net civilian migration, this is not the only component. Changes to the non-civilian population and definitional differences are also included.

Table 4.1 Live births, stillbirths, total births, deaths, infant and perinatal mortality during 2001 and conceptions during 2000

Area	Estimated resident population at 30 June 2001 (thousands)	Live births							Conceptions in 2000[5]			
		Male	Female	Total	TFR	% outside marriage		Proportion under 2,500 grams[2]	All ages	Under 18	All ages conception rate per 1,000 women aged 15-44	Under 18 conception rate per 1,000 women aged 15-17
						All	% jointly registered, same address[1]					
UNITED KINGDOM	**58,838.0**	**342,709**	**326,414**	**669,123**	**1.63**	**40.1**	**62.4**	-	-	-	-	-
ENGLAND AND WALES	**52,084.5**	**304,635**	**289,999**	**594,634**	**1.64**	**40.0**	**63.2**	**7.6**	**766,955**	**41,348**	**71.0**	**44.1**
ENGLAND	**49,181.3**	**288,743**	**275,001**	**563,744**	**1.64**	**39.6**	**63.3**	**7.6**	**729,063**	**38,699**	**71.3**	**43.9**
North East	**2,516.5**	13,380	12,569	25,949	1.58	50.9	57.2	7.4	32,467	2,478	62.6	51.4
North West	**6,731.5**	38,555	36,646	75,201	1.67	46.3	58.3	7.9	95,361	6,206	68.9	47.9
Yorkshire and The Humber	**4,967.2**	28,634	26,991	55,625	1.67	43.3	64.8	8.1	68,537	4,406	67.4	48.1
East Midlands	**4,175.1**	22,997	21,645	44,642	1.62	42.6	67.9	7.5	55,233	3,198	65.5	42.9
West Midlands	**5,267.1**	30,896	29,922	60,818	1.74	40.3	62.0	8.6	77,919	4,862	73.2	49.0
East	**5,394.9**	30,729	29,361	60,090	1.67	35.9	71.0	6.7	74,492	3,331	68.9	35.3
London	**7,188.0**	53,157	51,005	104,162	1.62	34.6	53.1	8.1	153,533	6,041	87.8	50.8
South East	**8,006.9**	45,457	43,053	88,510	1.62	34.8	70.7	6.9	111,290	5,085	68.5	36.1
South West	**4,934.2**	24,938	23,809	48,747	1.59	39.3	71.1	6.7	60,231	3,092	64.0	36.3
WALES	**2,903.2**	**15,746**	**14,870**	**30,616**	**1.66**	**48.3**	**60.8**	**7.5**	**37,892**	**2,649**	**66.3**	**48.4**
Normal residence outside England and Wales	-	146	128	274	-	18.2	44.0	17.6	-	-	-	
SCOTLAND[3]	**5,064.2**	**26,786**	**25,741**	**52,527**	**1.49**	**43.3**	**63.2**	-	**62,213**	**3,735**	**57.6**	**40.2**
Normal residence outside Scotland	-	115	114	229	-	28.4	58.5	-	-	-	-	-
NORTHERN IRELAND[4]	**1,689.3**	**11,288**	**10,674**	**21,962**	**1.80**	**32.5**	**33.4**	-	-	-	-	-
Normal residence outside Northern Ireland	-	160	129	289	-	12.1	48.6	-	-	-	-	-

Note: See Appendix A for fuller descriptions and definitions of table headings.

1 Number of live births outside marriage which are registered by both parents who gave the same address of usual residence, expressed as a percentage of all live births outside marriage.
2 Number of live births under 2,500 grams as a percentage of all live births for which the birthweight is known.
3 Includes births to women normally resident outside Scotland. Includes deaths of persons normally resident outside Scotland.
4 Excludes births to women normally resident outside Northern Ireland. Includes deaths of persons normally resident outside Northern Ireland.
5 The data shown in this table for Scotland may differ from that published elsewhere as these exclude miscarriages and other outcomes of pregnancy and are calculated using a different denominator.

United Kingdom by countries and, within England, Government Office Regions

Deaths				Deaths under 1 year				Stillbirth rate	Perinatal mortality rate	Infant mortality rate	Area
Male	Female	Total	SMR	Stillbirths	Under 1 week	Under 4 weeks	Under 1 year				
286,757	**315,511**	**602,268**	**100**	**3,572**	**1,821**	**2,434**	**3,664**	**5.3**	**8.0**	**5.5**	**UNITED KINGDOM**
252,426	**277,947**	**530,373**	**99**	**3,159**	**1,598**	**2,137**	**3,240**	**5.3**	**8.0**	**5.4**	**ENGLAND AND WALES**
236,038	**260,056**	**496,094**	**98**	**2,991**	**1,519**	**2,017**	**3,040**	**5.3**	**8.0**	**5.4**	**ENGLAND**
13,582	14,549	28,131	**110**	136	67	92	139	5.2	7.8	5.4	North East
35,244	39,389	74,633	109	459	197	284	439	6.1	8.7	5.8	North West
24,729	27,038	51,767	101	287	133	177	306	5.1	7.5	5.5	Yorkshire and The Humber
20,373	22,137	42,510	99	241	112	150	220	5.4	7.9	4.9	East Midlands
26,036	27,829	53,865	101	337	217	269	389	5.5	9.1	6.4	West Midlands
25,264	28,023	53,287	92	291	136	176	271	4.8	7.1	4.5	East
28,308	29,858	58,166	97	620	314	428	637	5.9	8.9	6.1	London
36,882	42,566	79,448	91	409	201	261	376	4.6	6.9	4.2	South East
25,620	28,667	54,287	90	211	142	180	263	4.3	7.2	5.4	South West
15,589	**17,404**	**32,993**	**103**	**155**	**76**	**106**	**164**	**5.0**	**7.5**	**5.4**	**WALES**
799	487	1,286	..	13	3	14	36	..	..	..	Normal residence outside England and Wales
27,324	**30,058**	**57,382**	115	**301**	**148**	**199**	**290**	**5.7**	**8.5**	**5.5**	**SCOTLAND**[3]
270	170	440	..	2	1	2	3	..	..	..	Normal residence outside Scotland
7,007	**7,506**	**14,513**	**103**	**112**	**75**	**98**	**134**	**5.1**	**8.4**	**6.0**	**NORTHERN IRELAND**[4]
42	36	78	..	-	3	3	3	..	..	..	Normal residence outside Northern Ireland

Map 3 Total fertility rates in 2001: English, Welsh and Scottish counties/unitary authorities, Northern Irish district council areas

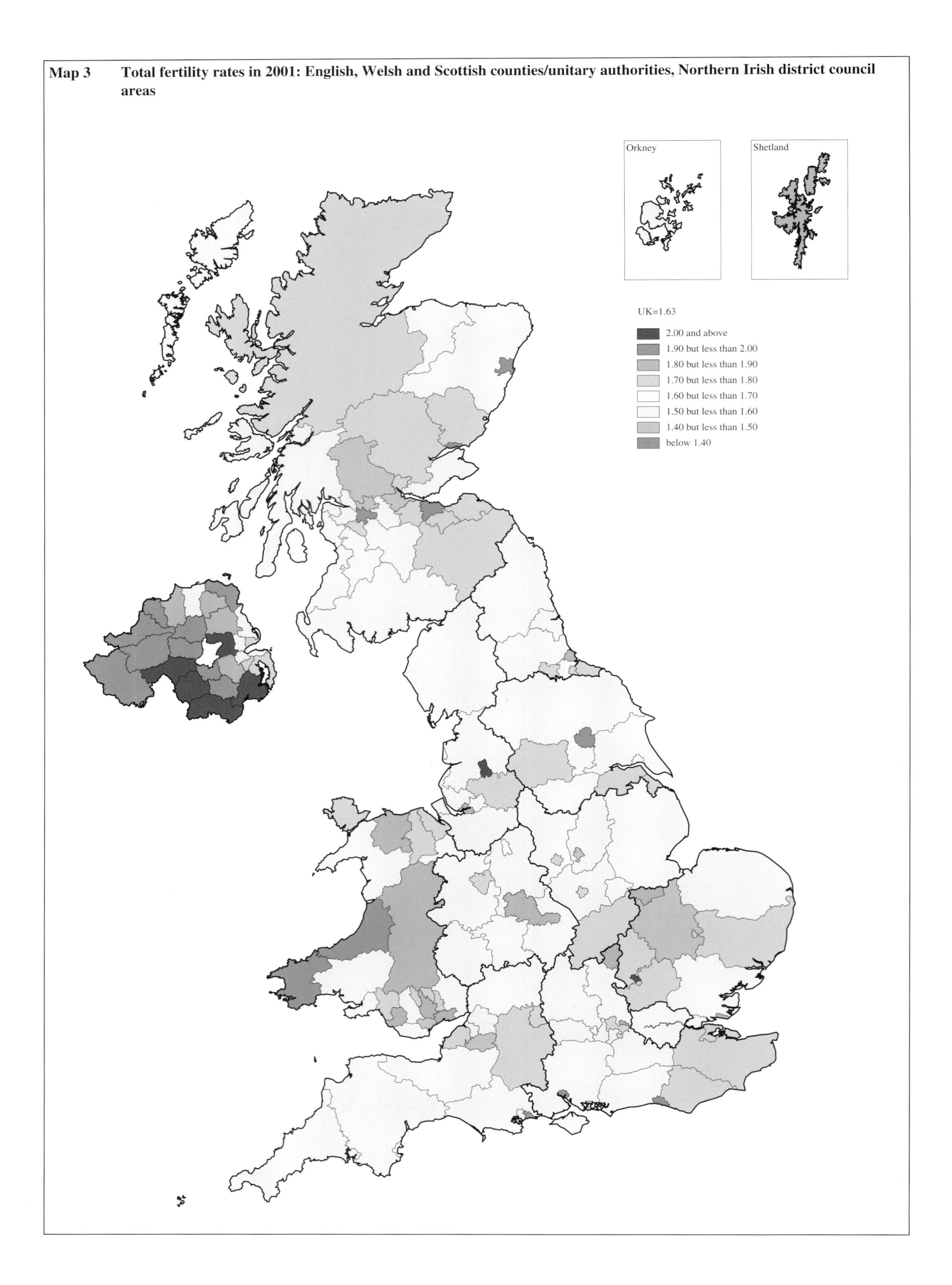

Map 4 **Standardised mortality ratios in 2001: English, Welsh and Scottish counties/unitary authorities, Northern Irish district council areas**

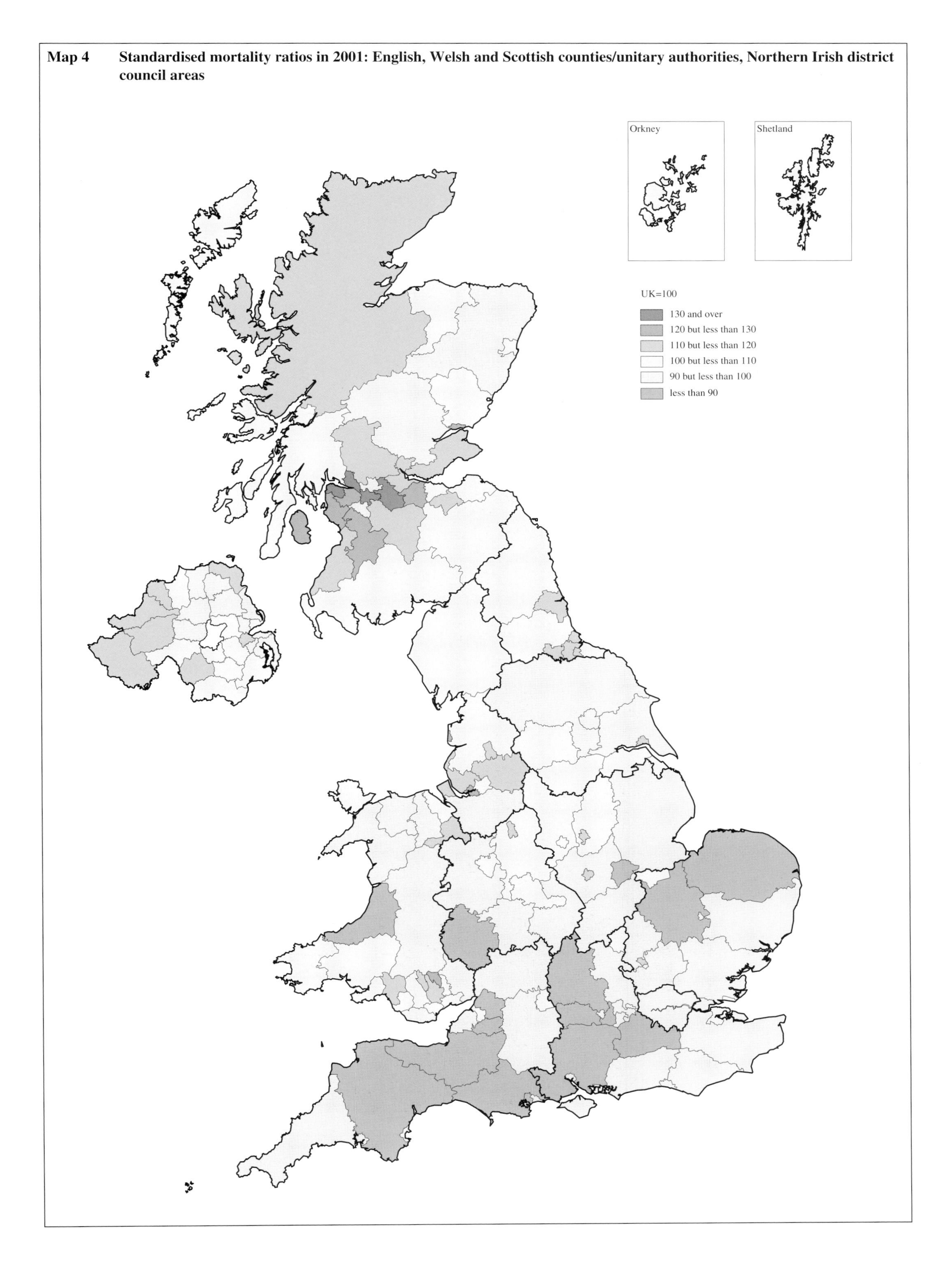

Table 4.2 Live births, stillbirths, total births, deaths, infant and perinatal mortality during 2001; and conceptions during 2000

Area	Estimated resident population at 30 June 2001 (thousands)	Live births							Conceptions in 2000			
		Male	Female	Total	TFR	% outside marriage		Proportion under 2,500 grams[2]	All ages	Under 18	All ages conception rate per 1,000 women aged 15-44	Under 18 conception rate per 1,000 women aged 15-17
						All	% jointly registered, same address[1]					
UNITED KINGDOM	**58,838.0**	**342,709**	**326,414**	**669,123**	**1.63**	**40.1**	**62.4**	-	-	-	-	-
ENGLAND AND WALES	**52,084.5**	**304,635**	**289,999**	**594,634**	**1.64**	**40.0**	**63.2**	**7.6**	**766,955**	**41,348**	**71.0**	**44.1**
ENGLAND	**49,181.3**	**288,743**	**275,001**	**563,744**	**1.64**	**39.6**	**63.3**	**7.6**	**729,063**	**38,699**	**71.3**	**43.9**
NORTH EAST	**2,516.5**	**13,380**	**12,569**	**25,949**	**1.58**	**50.9**	**57.2**	**7.4**	**32,467**	**2,478**	**62.6**	**51.4**
Darlington UA	97.9	566	512	1,078	1.76	50.1	63.7	7.2	1,299	94	65.9	52.2
Hartlepool UA	88.7	518	513	1,031	1.87	59.5	54.6	9.0	1,249	108	68.7	58.3
Middlesbrough UA	134.8	872	794	1,666	1.77	53.7	49.3	9.7	2,089	169	70.2	55.6
Redcar and Cleveland UA	139.2	744	688	1,432	1.70	56.8	57.4	8.1	1,797	162	65.6	57.2
Stockton-on-Tees UA	178.6	987	955	1,942	1.66	47.7	56.6	7.4	2,313	151	61.8	40.2
Durham	493.7	2,491	2,277	4,768	1.50	51.4	61.3	7.2	5,789	435	57.8	49.0
Chester-le-Street	53.7	270	253	523	1.54	44.6	58.8	8.6	608	42	54.9	43.3
Derwentside	85.2	446	426	872	1.67	53.1	56.8	8.7	960	62	57.0	41.4
Durham	87.8	362	353	715	1.17	43.5	63.7	4.9	903	57	45.8	36.3
Easington	94.0	485	448	933	1.60	60.0	58.8	6.8	1,202	98	63.5	55.3
Sedgefield	87.2	505	413	918	1.71	52.5	67.2	7.4	1,118	99	64.3	63.6
Teesdale	24.5	94	89	183	1.47	32.8	66.7	*4.9*	234	11	54.6	*28.4*
Wear Valley	61.4	329	295	624	1.74	54.6	62.2	7.7	764	66	63.8	59.1
Northumberland	307.4	1,433	1,448	2,881	1.65	42.3	66.9	6.1	3,545	238	60.9	40.9
Alnwick	31.1	146	144	290	1.75	39.0	76.1	*6.2*	334	11	59.3	*19.7*
Berwick-upon-Tweed	26.0	104	76	180	1.39	44.4	71.3	*7.2*	203	14	44.8	*31.7*
Blyth Valley	81.3	439	444	883	1.67	46.3	67.5	6.8	1,106	85	66.2	51.5
Castle Morpeth	49.0	189	189	378	1.58	30.7	62.1	*4.2*	475	29	55.0	30.7
Tynedale	58.9	240	282	522	1.71	34.5	75.6	4.6	614	27	58.4	23.4
Wansbeck	61.1	315	313	628	1.66	51.3	58.7	7.3	813	72	66.9	67.1
Tyne and Wear (Met County)	1,076.3	5,769	5,382	11,151	1.51	51.7	54.4	7.4	14,386	1,121	63.2	55.4
Gateshead	191.2	1,073	922	1,995	1.63	51.5	57.7	6.6	2,458	202	62.4	57.0
Newcastle upon Tyne	259.6	1,486	1,389	2,875	1.45	48.5	50.3	8.0	3,734	268	62.4	58.7
North Tyneside	192.0	1,005	944	1,949	1.59	46.6	60.0	6.1	2,566	190	66.9	55.3
South Tyneside	152.8	752	727	1,479	1.58	55.0	50.1	8.3	1,830	171	59.4	57.2
Sunderland	280.8	1,453	1,400	2,853	1.50	56.7	55.0	7.7	3,798	290	64.0	51.0
NORTH WEST	**6,731.5**	**38,555**	**36,646**	**75,201**	**1.67**	**46.3**	**58.3**	**7.9**	**95,361**	**6,206**	**68.9**	**47.9**
Blackburn with Darwen UA	137.6	1,052	1,031	2,083	2.21	37.7	58.8	10.7	2,473	161	85.5	55.0
Blackpool UA	142.3	684	713	1,397	1.66	61.0	62.4	7.6	1,881	171	69.9	70.5
Halton UA	118.2	721	729	1,450	1.82	57.3	53.8	7.2	1,894	144	74.3	54.8
Warrington UA	191.2	1,122	1,016	2,138	1.68	41.2	65.5	5.6	2,704	163	68.1	45.2
Cheshire	674.2	3,515	3,406	6,921	1.59	35.5	68.8	6.6	8,468	405	64.1	33.6
Chester	118.2	621	607	1,228	1.47	34.2	64.3	6.6	1,617	79	67.1	38.4
Congleton	90.8	429	423	852	1.46	29.2	77.5	4.5	1,025	38	58.5	23.9
Crewe and Nantwich	111.2	618	599	1,217	1.71	40.5	65.3	6.5	1,446	77	65.6	37.9
Ellesmere Port and Neston	81.6	402	414	816	1.59	42.0	66.8	9.0	1,018	51	62.7	34.2
Macclesfield	150.1	783	687	1,470	1.54	30.3	70.9	5.9	1,803	69	63.6	26.1
Vale Royal	122.3	662	676	1,338	1.78	38.0	71.3	7.4	1,559	91	65.5	40.7

Note: See Appendix A for fuller descriptions and definitions of table headings.

1 Number of live births outside marriage, which are registered by both parents who gave the same address of usual residence, expressed as a percentage of all live births outside marriage.

2 Number of live births under 2,500 grams as a percentage of all live births for which the birthweight is known.

United Kingdom by countries and, within England, Government Office Regions; counties/unitary authorities, districts and London boroughs/Northern Ireland council districts

Deaths				Deaths under 1 year				Stillbirth rate	Perinatal mortality rate	Infant mortality rate	Area
Male	Female	Total	SMR	Stillbirths	Under 1 week	Under 4 weeks	Under 1 year				
286,757	**315,511**	**602,268**	**100**	**3,572**	**1,821**	**2,434**	**3,664**	**5.3**	**8.0**	**5.5**	**UNITED KINGDOM**
252,426	**277,947**	**530,373**	**99**	**3,159**	**1,598**	**2,137**	**3,240**	**5.3**	**8.0**	**5.4**	**ENGLAND AND WALES**
236,038	**260,056**	**496,094**	**98**	**2,991**	**1,519**	**2,017**	**3,040**	**5.3**	**8.0**	**5.4**	**ENGLAND**
13,582	**14,549**	**28,131**	**110**	**136**	**67**	**92**	**139**	**5.2**	**7.8**	**5.4**	**NORTH EAST**
558	632	1,190	112	5	3	3	7	*4.6*	*7.4*	*6.5*	**Darlington UA**
500	486	986	116	3	1	2	5	*2.9*	*3.9*	*4.8*	**Hartlepool UA**
675	726	1,401	116	10	10	12	13	*6.0*	11.9	*7.8*	**Middlesbrough UA**
687	795	1,482	103	7	4	7	7	*4.9*	*7.6*	*4.9*	**Redcar and Cleveland UA**
872	907	1,779	111	14	6	8	13	*7.2*	10.2	*6.7*	**Stockton-on-Tees UA**
2,645	2,845	5,490	109	18	13	18	32	*3.8*	6.5	6.7	**Durham**
236	302	538	104	4	-	-	-	*7.6*	*7.6*	-	Chester-le-Street
509	552	1,061	115	-	1	2	5	-	*1.1*	*5.7*	Derwentside
394	430	824	103	1	4	5	8	*1.4*	*7.0*	*11.2*	Durham
531	526	1,057	112	3	2	2	5	*3.2*	*5.3*	*5.4*	Easington
478	498	976	114	8	4	6	7	*8.6*	*13.0*	*7.6*	Sedgefield
136	154	290	98	-	1	2	2	-	*5.5*	*10.9*	Teesdale
361	383	744	108	2	1	1	5	*3.2*	*4.8*	*8.0*	Wear Valley
1,673	1,796	3,469	103	19	3	6	10	*6.6*	7.6	*3.5*	**Northumberland**
162	160	322	85	3	-	-	-	*10.2*	*10.2*	-	Alnwick
149	184	333	93	1	-	-	1	*5.5*	*5.5*	*5.6*	Berwick-upon-Tweed
414	403	817	111	6	1	2	2	*6.7*	*7.9*	*2.3*	Blyth Valley
255	297	552	95	2	-	-	-	*5.3*	*5.3*	-	Castle Morpeth
310	369	679	99	2	-	1	2	*3.8*	*3.8*	*3.8*	Tynedale
383	383	766	120	5	2	3	5	*7.9*	*11.1*	*8.0*	Wansbeck
5,972	6,362	12,334	113	60	27	36	52	5.4	7.8	4.7	**Tyne and Wear (Met County)**
1,114	1,200	2,314	116	15	3	3	6	*7.5*	*9.0*	*3.0*	Gateshead
1,412	1,519	2,931	112	13	8	12	13	*4.5*	7.3	*4.5*	Newcastle upon Tyne
999	1,244	2,243	106	6	3	3	9	*3.1*	*4.6*	*4.6*	North Tyneside
908	844	1,752	108	10	6	9	12	*6.7*	*10.7*	*8.1*	South Tyneside
1,539	1,555	3,094	119	16	7	9	12	*5.6*	8.0	*4.2*	Sunderland
35,244	**39,389**	**74,633**	**109**	**459**	**197**	**284**	**439**	**6.1**	**8.7**	**5.8**	**NORTH WEST**
604	738	1,342	114	15	5	5	11	*7.1*	9.5	*5.3*	**Blackburn with Darwen UA**
1,046	1,093	2,139	120	13	4	5	8	*9.2*	*12.1*	*5.7*	**Blackpool UA**
603	613	1,216	122	7	4	4	7	*4.8*	*7.5*	*4.8*	**Halton UA**
870	958	1,828	105	9	5	6	8	*4.2*	*6.5*	*3.7*	**Warrington UA**
3,297	3,676	6,973	97	47	9	19	27	6.7	8.0	3.9	**Cheshire**
589	660	1,249	95	15	1	3	4	*12.1*	*12.9*	*3.3*	Chester
416	447	863	93	3	1	3	4	*3.5*	*4.7*	*4.7*	Congleton
570	637	1,207	106	6	2	5	7	*4.9*	*6.5*	*5.8*	Crewe and Nantwich
377	391	768	94	5	1	2	4	*6.1*	*7.3*	*4.9*	Ellesmere Port and Neston
757	907	1,664	94	12	3	3	4	*8.1*	*10.1*	*2.7*	Macclesfield
588	634	1,222	100	6	1	3	4	*4.5*	*5.2*	*3.0*	Vale Royal

Table 4.2 - *continued*

Area	Estimated resident population at 30 June 2001 (thousands)	Live births: Male	Female	Total	TFR	% outside marriage: All	% jointly registered, same address[1]	Proportion under 2,500 grams[2]	Conceptions in 2000: All ages	Under 18	All ages conception rate per 1,000 women aged 15-44	Under 18 conception rate per 1,000 women aged 15-17
Cumbria	487.8	2,335	2,300	4,635	1.61	42.5	66.9	6.8	5,701	360	61.7	41.0
Allerdale	93.5	438	468	906	1.66	40.2	62.1	7.0	1,073	76	61.6	46.3
Barrow-in-Furness	72.0	411	369	780	1.78	53.2	65.3	7.8	929	70	65.9	51.1
Carlisle	100.8	466	454	920	1.44	43.9	66.8	6.7	1,279	99	63.7	53.0
Copeland	69.2	347	317	664	1.62	47.7	60.6	5.1	797	64	58.7	49.7
Eden	49.9	234	237	471	1.67	35.9	81.7	7.9	569	14	63.2	*16.6*
South Lakeland	102.4	439	455	894	1.56	33.7	73.4	6.7	1,054	37	57.8	20.9
Greater Manchester (Met County)	2,482.8	15,387	14,577	29,964	1.71	46.3	58.7	8.3	38,342	2,479	72.9	52.1
Bolton	261.3	1,678	1,575	3,253	1.86	44.2	60.1	8.9	3,953	272	74.0	55.5
Bury	180.7	1,061	1,002	2,063	1.75	40.0	63.9	8.2	2,557	145	69.4	43.0
Manchester	392.9	2,809	2,687	5,496	1.62	52.9	47.1	8.2	7,721	524	80.7	70.2
Oldham	217.5	1,600	1,447	3,047	2.11	40.0	59.1	10.8	3,594	239	79.8	53.6
Rochdale	205.2	1,352	1,205	2,557	1.89	46.6	60.1	9.6	3,178	241	74.7	57.4
Salford	215.9	1,262	1,273	2,535	1.67	54.8	56.5	8.4	3,329	219	73.7	52.9
Stockport	284.6	1,509	1,394	2,903	1.56	40.8	65.4	6.8	3,757	173	65.7	32.7
Tameside	213.1	1,299	1,180	2,479	1.75	51.1	63.6	7.5	3,114	218	70.2	54.1
Trafford	210.2	1,165	1,202	2,367	1.65	36.3	61.8	6.5	2,954	138	68.0	33.5
Wigan	301.5	1,652	1,612	3,264	1.63	49.1	65.5	8.0	4,185	310	67.6	55.6
Lancashire	1,135.8	6,304	5,860	12,164	1.67	42.2	65.5	8.4	15,052	1,008	66.5	46.0
Burnley	89.5	558	498	1,056	1.83	46.5	63.3	9.3	1,311	92	70.8	49.1
Chorley	100.6	525	487	1,012	1.60	37.5	68.2	7.5	1,257	59	63.4	31.2
Fylde	73.4	335	301	636	1.67	38.5	69.8	6.4	793	41	63.6	35.1
Hyndburn	81.5	505	514	1,019	1.96	43.8	64.1	10.4	1,233	88	75.5	58.2
Lancaster	134.1	661	613	1,274	1.39	48.4	63.3	6.8	1,631	137	56.9	53.2
Pendle	89.3	547	536	1,083	1.93	36.2	66.3	10.2	1,400	110	79.1	58.7
Preston	129.6	876	783	1,659	1.67	43.9	56.2	10.3	2,074	142	70.9	53.9
Ribble Valley	54.1	263	240	503	1.57	32.8	80.0	5.8	611	19	61.9	*19.2*
Rossendale	65.7	383	381	764	1.84	46.9	69.6	7.6	949	70	71.6	56.5
South Ribble	103.9	575	543	1,118	1.71	42.1	71.3	7.9	1,247	72	60.4	36.1
West Lancashire	108.5	603	519	1,122	1.65	43.1	62.0	7.9	1,363	89	63.4	40.7
Wyre	105.8	473	445	918	1.60	39.3	72.0	7.7	1,183	89	64.4	45.5
Merseyside (Met County)	1,361.7	7,435	7,014	14,449	1.57	55.7	46.9	7.8	18,846	1,315	65.7	47.5
Knowsley	150.5	928	830	1,758	1.73	63.8	40.4	8.8	2,265	167	68.1	49.7
Liverpool	439.2	2,545	2,370	4,915	1.47	61.3	42.3	8.4	6,750	443	66.2	49.3
St Helens	176.8	940	888	1,828	1.60	51.6	55.0	7.8	2,241	176	63.0	51.2
Sefton	282.9	1,324	1,316	2,640	1.57	48.4	54.1	7.6	3,336	220	61.3	39.9
Wirral	312.2	1,698	1,610	3,308	1.74	51.4	49.7	6.7	4,254	309	69.0	48.3
YORKSHIRE AND THE HUMBER	**4,967.2**	**28,634**	**26,991**	**55,625**	**1.67**	**43.3**	**64.8**	**8.1**	**68,537**	**4,406**	**67.4**	**48.1**
East Riding of Yorkshire UA	314.8	1,434	1,358	2,792	1.60	39.7	72.5	6.9	3,406	174	59.8	30.9
Kingston upon Hull, City of UA	243.4	1,494	1,363	2,857	1.62	63.0	62.7	8.7	3,715	350	70.0	76.1
North East Lincolnshire UA	158.0	843	798	1,641	1.71	58.5	65.5	7.7	2,185	202	69.6	63.9
North Lincolnshire UA	153.0	819	780	1,599	1.79	48.2	64.7	8.0	1,999	128	67.8	45.2
York UA	181.3	946	883	1,829	1.36	42.0	64.4	6.5	2,219	100	57.3	33.0
North Yorkshire	570.1	2,851	2,581	5,432	1.64	31.8	71.1	5.7	6,647	304	63.4	30.5
Craven	53.7	233	229	462	1.59	30.5	77.3	5.2	595	23	64.0	24.0
Hambleton	84.2	422	377	799	1.70	27.8	70.3	4.1	929	42	61.3	28.9

Note: See Appendix A for fuller descriptions and definitions of table headings.

1 Number of live births outside marriage, which are registered by both parents who gave the same address of usual residence, expressed as a percentage of all live births outside marriage.

2 Number of live births under 2,500 grams as a percentage of all live births for which the birthweight is known.

United Kingdom by countries and, within England, Government Office Regions; counties/unitary authorities, districts and London boroughs/Northern Ireland council districts

Deaths				Deaths under 1 year				Stillbirth rate	Perinatal mortality rate	Infant mortality rate	Area
Male	Female	Total	SMR	Stillbirths	Under 1 week	Under 4 weeks	Under 1 year				
2,696	2,939	5,635	100	22	11	15	22	4.7	7.1	4.7	**Cumbria**
540	590	1,130	106	4	1	2	2	*4.4*	*5.5*	*2.2*	Allerdale
431	396	827	107	2	3	4	5	*2.6*	*6.4*	*6.4*	Barrow-in-Furness
559	628	1,187	105	5	-	-	2	*5.4*	*5.4*	*2.2*	Carlisle
361	358	719	103	2	1	3	4	*3.0*	*4.5*	*6.0*	Copeland
232	266	498	84	3	2	2	2	*6.3*	*10.5*	*4.2*	Eden
573	701	1,274	91	6	4	4	7	*6.7*	*11.1*	*7.8*	South Lakeland
12,783	14,102	26,885	114	197	83	119	196	6.5	9.3	6.5	**Greater Manchester (Met County)**
1,284	1,430	2,714	108	22	4	8	17	6.7	7.9	*5.2*	Bolton
854	1,018	1,872	108	18	5	5	8	*8.6*	11.1	*3.9*	Bury
2,251	2,222	4,473	133	44	16	26	53	7.9	10.8	9.6	Manchester
1,053	1,168	2,221	109	20	6	9	16	6.5	8.5	*5.3*	Oldham
1,009	1,164	2,173	116	18	12	16	20	*7.0*	11.7	7.8	Rochdale
1,249	1,393	2,642	120	17	9	10	14	*6.7*	10.2	*5.5*	Salford
1,391	1,549	2,940	98	14	12	14	21	*4.8*	8.9	7.2	Stockport
1,166	1,229	2,395	116	14	2	7	15	*5.6*	*6.4*	*6.1*	Tameside
1,005	1,132	2,137	98	10	5	8	11	*4.2*	*6.3*	*4.6*	Trafford
1,521	1,797	3,318	122	20	12	16	21	6.1	9.7	6.4	Wigan
5,857	6,788	12,645	105	65	36	57	79	5.3	8.3	6.5	**Lancashire**
420	551	971	110	7	1	2	6	*6.6*	*7.5*	*5.7*	Burnley
441	552	993	105	4	1	2	3	*3.9*	*4.9*	*3.0*	Chorley
462	573	1,035	96	6	1	1	1	*9.3*	*10.9*	*1.6*	Fylde
421	489	910	113	5	2	4	5	*4.9*	*6.8*	*4.9*	Hyndburn
777	789	1,566	101	8	6	12	13	*6.2*	*10.9*	*10.2*	Lancaster
431	497	928	103	12	5	10	16	*11.0*	*15.5*	*14.8*	Pendle
650	732	1,382	116	11	7	9	11	6.6	*10.8*	6.6	Preston
268	326	594	99	1	1	1	3	*2.0*	*4.0*	*6.0*	Ribble Valley
309	386	695	115	6	5	6	7	*7.8*	*14.3*	*9.2*	Rossendale
497	534	1,031	101	3	3	4	7	*2.7*	*5.4*	*6.3*	South Ribble
562	630	1,192	111	1	3	4	5	*0.9*	*3.6*	*4.5*	West Lancashire
619	729	1,348	94	1	1	2	2	*1.1*	2.2	2.2	Wyre
7,488	8,482	15,970	115	84	40	54	81	5.8	8.5	5.6	**Merseyside (Met County)**
743	783	1,526	122	11	4	5	12	*6.2*	*8.5*	*6.8*	Knowsley
2,457	2,718	5,175	126	26	16	23	33	5.3	8.5	6.7	Liverpool
911	1,029	1,940	115	11	8	8	13	*6.0*	*10.3*	*7.1*	St Helens
1,607	1,943	3,550	107	11	7	7	9	*4.1*	*6.8*	*3.4*	Sefton
1,770	2,009	3,779	106	25	5	11	14	7.5	9.0	*4.2*	Wirral
24,729	**27,038**	**51,767**	**101**	**287**	**133**	**177**	**306**	**5.1**	**7.5**	**5.5**	**YORKSHIRE AND THE HUMBER**
1,584	1,794	3,378	92	8	9	13	16	*2.9*	*6.1*	*5.7*	**East Riding of Yorkshire UA**
1,286	1,267	2,553	110	12	3	7	14	*4.2*	*5.2*	*4.9*	**Kingston upon Hull, City of UA**
846	828	1,674	101	10	6	7	12	*6.1*	*9.7*	*7.3*	**North East Lincolnshire UA**
802	839	1,641	102	8	4	5	8	*5.0*	*7.5*	*5.0*	**North Lincolnshire UA**
873	945	1,818	92	7	5	6	7	*3.8*	*6.5*	*3.8*	**York UA**
2,869	3,217	6,086	90	19	5	8	10	*3.5*	4.4	*1.8*	**North Yorkshire**
261	369	630	88	3	1	2	2	*6.5*	*8.6*	*4.3*	Craven
397	420	817	87	3	-	1	1	*3.7*	*3.7*	*1.3*	Hambleton

Table 4.2 - *continued*

Area	Estimated resident population at 30 June 2001 (thousands)	Live births							Conceptions in 2000			
		Male	Female	Total	TFR	% outside marriage		Proportion under 2,500 grams[2]	All ages	Under 18	All ages conception rate per 1,000 women aged 15-44	Under 18 conception rate per 1,000 women aged 15-17
						All	% jointly registered, same address[1]					
Harrogate	151.5	807	726	1,533	1.58	27.4	71.2	5.8	1,857	60	62.8	22.1
Richmondshire	47.1	277	251	528	1.87	21.6	75.4	6.3	647	16	76.6	22.2
Ryedale	50.9	209	196	405	1.51	31.1	74.6	*4.7*	490	22	58.7	29.6
Scarborough	106.2	489	427	916	1.60	48.4	65.5	6.1	1,197	100	63.4	52.7
Selby	76.6	414	375	789	1.67	33.2	74.0	6.7	932	41	61.9	27.7
South Yorkshire (Met County)	1,266.5	7,135	6,725	13,860	1.63	48.5	63.3	8.0	17,430	1,255	66.9	55.3
Barnsley	218.1	1,127	1,099	2,226	1.64	53.1	65.1	7.3	2,762	207	62.7	52.5
Doncaster	286.9	1,640	1,585	3,225	1.85	52.3	65.0	8.3	4,054	386	70.9	70.0
Rotherham	248.4	1,429	1,301	2,730	1.74	48.8	60.5	7.7	3,375	238	67.7	51.2
Sheffield	513.1	2,939	2,740	5,679	1.53	44.3	62.9	8.2	7,239	424	66.2	49.4
West Yorkshire (Met County)	2,080.2	13,112	12,503	25,615	1.73	40.0	64.2	8.8	30,936	1,893	70.1	47.7
Bradford	467.9	3,743	3,462	7,205	2.18	32.8	61.3	9.6	8,354	499	84.6	49.6
Calderdale	192.4	1,160	1,106	2,266	1.86	43.6	65.1	9.2	2,653	158	68.1	43.9
Kirklees	388.9	2,552	2,480	5,032	1.87	36.7	66.4	9.0	5,972	317	73.9	43.4
Leeds	715.5	4,034	3,797	7,831	1.43	44.0	63.7	8.1	9,941	631	62.8	49.1
Wakefield	315.4	1,623	1,658	3,281	1.59	48.7	66.5	8.2	4,016	288	62.0	48.9
EAST MIDLANDS	**4,175.1**	**22,997**	**21,645**	**44,642**	**1.62**	**42.6**	**67.9**	**7.5**	**55,233**	**3,198**	**65.5**	**42.9**
Derby UA	221.7	1,442	1,350	2,792	1.75	44.4	63.2	8.2	3,433	206	72.2	49.5
Leicester UA	279.8	2,070	1,915	3,985	1.72	41.9	56.0	9.6	5,138	329	77.5	57.4
Nottingham UA	267.0	1,703	1,576	3,279	1.48	58.1	51.2	9.7	4,249	348	67.1	71.8
Rutland UA	34.6	161	138	299	1.67	29.8	85.4	7.4	373	21	60.1	29.2
Derbyshire	734.9	3,794	3,485	7,279	1.61	43.1	73.7	6.9	8,858	479	62.1	37.8
Amber Valley	116.6	649	537	1,186	1.64	41.5	74.2	6.6	1,419	65	63.3	33.3
Bolsover	71.9	382	361	743	1.70	49.7	69.6	7.9	934	59	66.4	46.7
Chesterfield	98.8	500	427	927	1.49	50.1	67.7	6.8	1,209	70	62.0	43.8
Derbyshire Dales	69.4	316	304	620	1.68	32.6	81.7	7.6	712	24	59.2	20.0
Erewash	110.1	594	568	1,162	1.61	47.4	73.5	8.1	1,404	82	63.0	43.0
High Peak	89.4	439	459	898	1.60	41.0	74.5	5.2	1,130	73	63.8	44.5
North East Derbyshire	96.9	420	416	836	1.52	43.2	75.9	6.5	1,004	65	56.0	38.7
South Derbyshire	81.7	494	413	907	1.66	36.1	78.6	6.9	1,046	41	62.6	28.5
Leicestershire	610.3	3,201	3,018	6,219	1.56	33.8	76.5	6.9	7,663	313	63.4	29.4
Blaby	90.4	493	453	946	1.55	31.9	76.2	7.6	1,132	35	63.9	23.5
Charnwood	153.6	818	740	1,558	1.48	36.0	72.5	7.2	1,961	87	61.0	31.7
Harborough	76.8	444	373	817	1.70	26.9	83.6	4.5	970	35	66.8	26.9
Hinckley and Bosworth	100.2	489	507	996	1.57	36.7	80.3	8.2	1,254	56	64.4	31.8
Melton	47.9	226	223	449	1.53	33.9	75.7	7.3	550	28	59.3	32.9
North West Leicestershire	85.7	474	472	946	1.72	37.0	76.9	6.8	1,137	45	68.3	31.4
Oadby and Wigston	55.8	257	250	507	1.46	30.4	72.7	5.9	659	27	59.6	25.7
Lincolnshire	647.6	3,128	2,892	6,020	1.61	43.2	72.2	6.9	7,428	446	62.4	40.4
Boston	55.8	286	253	539	1.75	51.9	68.2	8.0	692	48	68.8	52.0
East Lindsey	130.7	517	518	1,035	1.63	45.7	72.9	6.8	1,309	85	61.1	41.8
Lincoln	85.6	504	450	954	1.46	54.3	65.3	7.8	1,245	111	65.2	73.8
North Kesteven	94.4	454	452	906	1.64	30.2	77.4	6.4	1,068	43	62.4	28.3
South Holland	76.7	354	327	681	1.67	44.3	75.5	4.4	833	53	64.0	44.3

Note: See Appendix A for fuller descriptions and definitions of table headings.

1 Number of live births outside marriage, which are registered by both parents who gave the same address of usual residence, expressed as a percentage of all live births outside marriage.

2 Number of live births under 2,500 grams as a percentage of all live births for which the birthweight is known.

United Kingdom by countries and, within England, Government Office Regions; counties/unitary authorities, districts and London boroughs/Northern Ireland council districts

Deaths				Deaths under 1 year				Stillbirth rate	Perinatal mortality rate	Infant mortality rate	Area
Male	Female	Total	SMR	Stillbirths	Under 1 week	Under 4 weeks	Under 1 year				
715	845	1,560	88	5	3	3	3	*3.3*	*5.2*	*2.0*	Harrogate
219	212	431	92	1	1	2	2	*1.9*	*3.8*	*3.8*	Richmondshire
264	264	528	79	1	-	-	-	*2.5*	*2.5*	-	Ryedale
647	748	1,395	94	4	-	-	2	*4.3*	*4.3*	*2.2*	Scarborough
366	359	725	98	2	-	-	-	*2.5*	*2.5*	-	Selby
6,561	7,097	13,658	105	76	35	44	73	5.5	8.0	5.3	**South Yorkshire (Met County)**
1,207	1,281	2,488	113	8	5	5	10	*3.6*	*5.8*	*4.5*	Barnsley
1,443	1,559	3,002	105	18	3	6	14	*5.6*	6.5	*4.3*	Doncaster
1,245	1,322	2,567	106	11	8	9	15	*4.0*	*6.9*	*5.5*	Rotherham
2,666	2,935	5,601	102	39	19	24	34	6.8	10.1	6.0	Sheffield
9,908	11,051	20,959	104	147	66	87	166	5.7	8.3	6.5	**West Yorkshire (Met County)**
2,213	2,502	4,715	108	55	23	30	63	7.6	10.7	8.7	Bradford
922	1,089	2,011	101	15	6	8	13	*6.6*	9.2	*5.7*	Calderdale
1,821	2,113	3,934	106	18	15	16	30	*3.6*	6.5	6.0	Kirklees
3,321	3,628	6,949	98	48	14	22	39	6.1	7.9	5.0	Leeds
1,631	1,719	3,350	110	11	8	11	21	*3.3*	*5.8*	6.4	Wakefield
20,373	**22,137**	**42,510**	**99**	**241**	**112**	**150**	**220**	**5.4**	**7.9**	**4.9**	**EAST MIDLANDS**
1,107	1,195	2,302	102	22	13	16	20	7.8	12.4	7.2	**Derby UA**
1,332	1,437	2,769	110	32	16	21	27	8.0	11.9	6.8	**Leicester UA**
1,405	1,407	2,812	115	15	14	20	31	*4.6*	8.8	9.5	**Nottingham UA**
135	158	293	78	1	2	2	3	*3.3*	*10.0*	*10.0*	**Rutland UA**
3,800	4,186	7,986	101	47	12	19	28	6.4	8.1	3.8	**Derbyshire**
594	643	1,237	97	11	3	4	6	*9.2*	*11.7*	*5.1*	Amber Valley
387	443	830	107	5	-	1	3	*6.7*	*6.7*	*4.0*	Bolsover
552	634	1,186	107	5	3	5	6	*5.4*	*8.6*	*6.5*	Chesterfield
383	455	838	97	1	1	1	1	*1.6*	*3.2*	*1.6*	Derbyshire Dales
557	597	1,154	102	6	-	1	1	*5.1*	*5.1*	*0.9*	Erewash
408	455	863	94	7	-	-	3	*7.7*	*7.7*	*3.3*	High Peak
536	561	1,097	102	9	2	2	3	*10.7*	*13.0*	*3.6*	North East Derbyshire
383	398	781	102	3	3	5	5	*3.3*	*6.6*	*5.5*	South Derbyshire
2,590	2,949	5,539	90	32	15	20	25	5.1	7.5	4.0	**Leicestershire**
320	340	660	77	10	2	4	4	*10.5*	*12.6*	*4.2*	Blaby
651	792	1,443	97	4	4	4	6	*2.6*	*5.1*	*3.9*	Charnwood
349	390	739	95	1	2	3	3	*1.2*	*3.7*	*3.7*	Harborough
429	497	926	89	7	2	3	4	*7.0*	*9.0*	*4.0*	Hinckley and Bosworth
213	233	446	89	4	1	1	1	*8.8*	*11.0*	*2.2*	Melton
392	424	816	93	4	1	2	3	*4.2*	*5.3*	*3.2*	North West Leicestershire
236	273	509	87	2	3	3	4	*3.9*	*9.8*	*7.9*	Oadby and Wigston
3,588	3,708	7,296	96	28	10	15	26	4.6	6.3	4.3	**Lincolnshire**
318	370	688	100	5	-	-	2	*9.2*	*9.2*	*3.7*	Boston
860	839	1,699	96	5	-	1	4	*4.8*	*4.8*	*3.9*	East Lindsey
431	406	837	98	6	4	5	9	*6.3*	*10.4*	*9.4*	Lincoln
452	562	1,014	95	-	-	2	2	-	-	*2.2*	North Kesteven
501	494	995	100	5	-	1	1	*7.3*	*7.3*	*1.5*	South Holland

Table 4.2 - *continued*

Area	Estimated resident population at 30 June 2001 (thousands)	Live births							Conceptions in 2000			
		Male	Female	Total	TFR	% outside marriage		Proportion under 2,500 grams[2]	All ages	Under 18	All ages conception rate per 1,000 women aged 15-44	Under 18 conception rate per 1,000 women aged 15-17
						All	% jointly registered, same address[1]					
South Kesteven	124.9	654	604	1,258	1.68	39.5	74.4	6.7	1,471	67	60.3	28.6
West Lindsey	79.6	359	288	647	1.57	39.9	75.2	8.3	810	39	57.7	25.9
Northamptonshire	630.4	3,764	3,620	7,384	1.75	41.5	68.9	7.2	9,089	525	70.1	45.5
Corby	53.2	305	325	630	1.88	57.9	61.4	7.5	765	78	68.8	69.3
Daventry	72.1	390	399	789	1.76	32.1	74.3	5.7	929	28	66.8	22.6
East Northamptonshire	76.8	418	413	831	1.71	35.9	75.2	6.5	1,065	68	71.6	46.1
Kettering	82.0	493	489	982	1.80	40.6	72.2	5.8	1,195	57	72.7	39.4
Northampton	194.4	1,279	1,200	2,479	1.68	46.3	65.9	8.0	3,155	184	72.1	51.8
South Northamptonshire	79.5	441	415	856	1.75	26.4	85.4	8.2	994	31	64.8	22.1
Wellingborough	72.6	438	379	817	1.78	45.5	63.4	7.1	986	79	69.0	60.7
Nottinghamshire	748.8	3,734	3,651	7,385	1.54	43.6	69.8	7.2	9,002	531	60.9	40.2
Ashfield	111.6	560	589	1,149	1.56	50.2	70.9	6.1	1,390	97	61.6	50.1
Bassetlaw	107.8	510	537	1,047	1.64	46.7	65.4	8.4	1,258	84	61.2	44.1
Broxtowe	107.5	519	485	1,004	1.39	38.5	67.7	7.0	1,244	64	57.7	35.7
Gedling	111.8	549	528	1,077	1.47	42.5	66.6	7.1	1,321	88	59.7	44.4
Mansfield	98.0	521	460	981	1.59	55.8	70.6	8.9	1,256	102	62.2	54.3
Newark and Sherwood	106.4	551	533	1,084	1.74	45.3	68.8	7.4	1,254	60	62.1	30.9
Rushcliffe	105.8	524	519	1,043	1.44	25.8	84.0	5.6	1,279	36	61.6	20.2
WEST MIDLANDS	**5,267.1**	**30,896**	**29,922**	**60,818**	**1.74**	**40.3**	**62.0**	**8.6**	**77,919**	**4,862**	**73.2**	**49.0**
Herefordshire, County of UA	174.9	783	808	1,591	1.63	38.2	70.3	6.7	1,921	117	61.4	39.0
Stoke-on-Trent UA	240.4	1,389	1,352	2,741	1.63	51.7	64.2	8.9	3,410	274	66.7	59.8
Telford and Wrekin UA	158.5	966	910	1,876	1.71	49.5	64.1	8.5	2,341	179	70.5	57.8
Shropshire	283.3	1,317	1,311	2,628	1.62	35.8	75.1	6.2	3,246	144	63.4	28.7
Bridgnorth	52.5	198	223	421	1.45	24.9	85.7	8.3	516	16	56.9	*17.6*
North Shropshire	57.2	285	273	558	1.75	35.1	75.5	6.1	675	30	66.6	29.7
Oswestry	37.3	156	156	312	1.36	41.0	72.7	*5.1*	377	20	52.9	27.8
Shrewsbury and Atcham	95.9	493	508	1,001	1.70	39.9	71.7	6.3	1,246	59	68.2	34.5
South Shropshire	40.4	185	151	336	1.76	33.3	79.5	*4.2*	432	19	65.6	*28.4*
Staffordshire	807.1	4,039	4,022	8,061	1.59	38.6	72.0	7.9	9,979	587	63.1	39.3
Cannock Chase	92.2	518	523	1,041	1.66	45.4	72.3	9.5	1,322	101	68.6	58.7
East Staffordshire	103.9	625	608	1,233	1.87	34.9	67.2	10.0	1,473	103	71.2	51.5
Lichfield	93.2	416	439	855	1.54	33.7	74.7	6.4	1,037	56	60.0	32.8
Newcastle-under-Lyme	122.0	563	577	1,140	1.41	45.3	71.3	8.2	1,395	92	55.9	41.6
South Staffordshire	105.9	454	436	890	1.45	32.0	75.8	7.5	1,123	58	56.8	29.6
Stafford	120.7	535	580	1,115	1.53	35.5	74.7	6.7	1,426	57	62.7	27.0
Staffordshire Moorlands	94.6	440	402	842	1.54	36.3	70.9	6.5	998	53	57.6	31.8
Tamworth	74.6	488	457	945	1.79	44.1	71.2	7.7	1,205	67	74.2	42.9
Warwickshire	506.2	2,642	2,612	5,254	1.60	36.2	68.0	7.3	6,914	373	70.4	41.9
North Warwickshire	61.8	339	296	635	1.65	39.5	70.1	7.7	819	45	66.7	39.8
Nuneaton and Bedworth	119.3	670	694	1,364	1.76	44.4	60.7	7.8	1,770	116	74.1	52.1
Rugby	87.5	478	490	968	1.74	39.2	73.4	8.2	1,285	80	76.3	49.3
Stratford-on-Avon	111.5	546	529	1,075	1.62	26.4	76.1	6.1	1,400	59	69.8	31.9
Warwick	126.1	609	603	1,212	1.35	31.4	66.8	6.8	1,640	73	65.2	35.1

Note: See Appendix A for fuller descriptions and definitions of table headings.

1 Number of live births outside marriage, which are registered by both parents who gave the same address of usual residence, expressed as a percentage of all live births outside marriage.

2 Number of live births under 2,500 grams as a percentage of all live births for which the birthweight is known.

United Kingdom by countries and, within England, Government Office Regions; counties/unitary authorities, districts and London boroughs/Northern Ireland council districts

Deaths				Deaths under 1 year				Stillbirth rate	Perinatal mortality rate	Infant mortality rate	Area
Male	Female	Total	SMR	Stillbirths	Under 1 week	Under 4 weeks	Under 1 year				
586	593	1,179	90	3	4	4	6	*2.4*	*5.6*	*4.8*	South Kesteven
440	444	884	97	4	2	2	2	*6.1*	*9.2*	*3.1*	West Lindsey
2,686	2,863	5,549	94	37	13	16	26	5.0	6.7	3.5	**Northamptonshire**
248	242	490	110	3	1	1	2	*4.7*	*6.3*	*3.2*	Corby
301	296	597	94	3	1	1	1	*3.8*	*5.1*	*1.3*	Daventry
340	384	724	94	3	1	1	1	*3.6*	*4.8*	*1.2*	East Northamptonshire
366	405	771	90	4	3	3	3	*4.1*	*7.1*	*3.1*	Kettering
808	880	1,688	96	17	3	6	13	6.8	8.0	5.2	Northampton
302	324	626	85	4	4	4	5	*4.7*	*9.3*	*5.8*	South Northamptonshire
321	332	653	93	3	-	-	1	*3.7*	*3.7*	*1.2*	Wellingborough
3,730	4,234	7,964	101	27	17	21	34	3.6	5.9	4.6	**Nottinghamshire**
616	646	1,262	113	5	3	3	7	*4.3*	*6.9*	*6.1*	Ashfield
597	607	1,204	109	6	5	7	11	*5.7*	*10.4*	*10.5*	Bassetlaw
509	583	1,092	97	6	2	3	6	*5.9*	*7.9*	*6.0*	Broxtowe
468	598	1,066	88	3	1	1	1	*2.8*	*3.7*	*0.9*	Gedling
523	552	1,075	108	-	1	1	1	-	*1.0*	*1.0*	Mansfield
545	651	1,196	102	5	2	2	4	*4.6*	*6.4*	*3.7*	Newark and Sherwood
472	597	1,069	95	2	3	4	4	*1.9*	*4.8*	*3.8*	Rushcliffe
26,036	**27,829**	**53,865**	**101**	**337**	**217**	**269**	**389**	**5.5**	**9.1**	**6.4**	**WEST MIDLANDS**
907	969	1,876	88	9	4	4	4	*5.6*	*8.1*	2.5	**Herefordshire, County of UA**
1,275	1,396	2,671	110	16	15	18	22	*5.8*	11.2	8.0	**Stoke-on-Trent UA**
640	686	1,326	103	13	10	10	14	*6.9*	12.2	*7.5*	**Telford and Wrekin UA**
1,425	1,605	3,030	92	15	9	11	13	*5.7*	9.1	*4.9*	**Shropshire**
275	272	547	96	1	4	5	5	*2.4*	*11.8*	*11.9*	Bridgnorth
277	329	606	92	4	1	1	2	*7.1*	*8.9*	*3.6*	North Shropshire
183	234	417	99	1	-	-	-	*3.2*	*3.2*	-	Oswestry
447	505	952	88	8	2	3	4	*7.9*	*9.9*	*4.0*	Shrewsbury and Atcham
243	265	508	91	1	2	2	2	*3.0*	*8.9*	*6.0*	South Shropshire
3,877	4,291	8,168	102	40	25	32	42	4.9	8.0	5.2	**Staffordshire**
416	385	801	103	5	6	6	7	*4.8*	*10.5*	*6.7*	Cannock Chase
577	586	1,163	113	9	3	4	7	*7.2*	*9.7*	*5.7*	East Staffordshire
404	568	972	103	4	-	2	3	*4.7*	*4.7*	*3.5*	Lichfield
602	672	1,274	99	3	3	3	4	*2.6*	*5.2*	*3.5*	Newcastle-under-Lyme
454	549	1,003	94	7	4	4	5	*7.8*	*12.3*	*5.6*	South Staffordshire
607	695	1,302	98	3	4	7	8	*2.7*	*6.3*	*7.2*	Stafford
502	536	1,038	100	3	4	4	6	*3.6*	*8.3*	*7.1*	Staffordshire Moorlands
315	300	615	113	6	1	2	2	*6.3*	*7.4*	*2.1*	Tamworth
2,418	2,716	5,134	97	18	9	15	24	*3.4*	5.1	4.6	**Warwickshire**
304	313	617	105	6	1	2	2	*9.4*	*10.9*	*3.1*	North Warwickshire
589	596	1,185	108	4	5	6	9	*2.9*	*6.6*	*6.6*	Nuneaton and Bedworth
426	473	899	97	1	1	1	1	*1.0*	*2.1*	*1.0*	Rugby
539	676	1,215	92	2	-	1	4	*1.9*	*1.9*	*3.7*	Stratford-on-Avon
560	658	1,218	90	5	2	5	8	*4.1*	*5.8*	*6.6*	Warwick

Table 4.2 - *continued*

Area	Estimated resident population at 30 June 2001 (thousands)	Live births: Male	Live births: Female	Live births: Total	TFR	% outside marriage: All	% outside marriage: % jointly registered, same address[1]	Proportion under 2,500 grams[2]	Conceptions in 2000: All ages	Conceptions in 2000: Under 18	All ages conception rate per 1,000 women aged 15-44	Under 18 conception rate per 1,000 women aged 15-17
West Midlands (Met County)	2,554.4	16,900	16,205	33,105	1.85	40.8	55.9	9.5	43,207	2,837	80.7	56.8
Birmingham	976.4	7,358	7,068	14,426	1.96	37.1	51.5	9.9	18,724	1,133	87.2	55.9
Coventry	300.7	1,848	1,766	3,614	1.68	45.2	60.3	8.8	5,134	374	80.3	64.4
Dudley	305.1	1,724	1,589	3,313	1.70	40.7	65.3	7.8	4,271	272	71.3	50.9
Sandwell	282.8	1,854	1,840	3,694	1.92	43.5	53.6	10.6	4,845	346	82.2	63.3
Solihull	199.6	1,038	921	1,959	1.63	38.8	62.4	6.9	2,519	120	66.5	32.2
Walsall	253.3	1,621	1,578	3,199	1.99	43.9	61.5	9.7	3,926	306	77.2	62.7
Wolverhampton	236.4	1,457	1,443	2,900	1.78	47.9	52.3	10.2	3,788	286	77.3	63.5
Worcestershire	542.2	2,860	2,702	5,562	1.62	37.6	69.9	7.6	6,901	351	65.4	36.2
Bromsgrove	87.9	418	383	801	1.53	30.7	79.7	5.6	1,027	41	63.9	27.3
Malvern Hills	72.2	317	276	593	1.65	39.3	73.4	7.0	769	47	62.7	31.1
Redditch	78.8	486	481	967	1.74	43.2	69.1	8.0	1,215	68	71.5	43.2
Worcester	93.4	591	571	1,162	1.59	36.1	64.0	9.0	1,473	79	70.3	49.6
Wychavon	113.1	549	527	1,076	1.59	35.4	72.2	6.9	1,259	47	60.5	25.2
Wyre Forest	96.9	499	464	963	1.59	41.1	66.4	8.4	1,158	69	63.1	41.6
EAST	**5,394.9**	**30,729**	**29,361**	**60,090**	**1.67**	**35.9**	**71.0**	**6.7**	**74,492**	**3,331**	**68.9**	**35.3**
Luton UA	184.3	1,463	1,391	2,854	2.01	30.6	57.6	9.7	3,660	170	87.7	44.0
Peterborough UA	156.5	1,037	1,021	2,058	1.84	41.3	62.4	7.8	2,522	147	74.8	47.4
Southend-on-Sea UA	160.4	968	936	1,904	1.83	48.1	65.5	6.2	2,462	126	77.0	47.9
Thurrock UA	143.2	934	863	1,797	1.66	46.7	68.1	5.8	2,248	133	72.0	52.2
Bedfordshire	382.1	2,291	2,190	4,481	1.70	32.2	70.2	7.4	5,558	237	70.7	34.5
Bedford	148.1	852	852	1,704	1.61	33.2	63.8	7.9	2,181	91	71.2	33.8
Mid Bedfordshire	121.3	750	668	1,418	1.69	27.8	78.4	6.1	1,720	65	69.2	31.5
South Bedfordshire	112.7	689	670	1,359	1.82	35.5	71.2	8.1	1,657	81	71.6	38.4
Cambridgeshire	553.6	3,026	2,832	5,858	1.48	32.1	74.1	6.9	7,230	281	62.0	29.9
Cambridge	108.8	563	501	1,064	1.19	34.6	69.6	7.8	1,503	52	53.2	34.3
East Cambridgeshire	73.4	417	342	759	1.57	28.2	78.5	6.6	903	33	62.1	25.9
Fenland	83.7	414	431	845	1.73	44.0	68.3	7.9	994	68	64.0	49.5
Huntingdonshire	157.2	937	877	1,814	1.73	31.9	75.8	6.9	2,167	89	66.1	30.7
South Cambridgeshire	130.5	695	681	1,376	1.60	25.2	79.5	5.6	1,663	39	65.1	16.7
Essex	1,312.7	7,232	6,851	14,083	1.65	36.5	72.2	6.2	17,585	785	68.6	34.9
Basildon	165.9	1,039	1,035	2,074	1.77	44.0	67.9	6.7	2,611	143	74.8	48.0
Braintree	132.5	803	702	1,505	1.71	32.4	74.6	7.0	1,785	72	68.2	33.4
Brentwood	68.5	345	331	676	1.55	28.0	75.1	5.9	820	24	62.7	20.5
Castle Point	86.7	429	406	835	1.64	39.3	77.7	6.3	1,069	53	67.3	34.4
Chelmsford	157.2	803	809	1,612	1.48	31.3	71.4	6.1	2,022	76	62.7	26.8
Colchester	156.0	835	821	1,656	1.50	34.9	71.1	5.7	2,084	115	64.6	43.3
Epping Forest	121.0	706	661	1,367	1.64	35.3	72.5	6.4	1,778	60	74.5	29.9
Harlow	78.9	562	529	1,091	1.84	45.0	70.7	6.5	1,383	80	79.1	53.1
Maldon	59.6	352	260	612	1.70	33.3	76.5	4.7	756	24	68.9	24.4
Rochford	78.7	395	418	813	1.67	30.8	77.2	6.2	1,015	37	68.7	27.4
Tendring	138.8	593	546	1,139	1.71	46.5	70.2	6.3	1,417	86	65.4	41.7
Uttlesford	69.0	370	333	703	1.68	25.7	77.3	5.6	845	15	64.9	*11.8*

Note: See Appendix A for fuller descriptions and definitions of table headings.

1 Number of live births outside marriage, which are registered by both parents who gave the same address of usual residence, expressed as a percentage of all live births outside marriage.
2 Number of live births under 2,500 grams as a percentage of all live births for which the birthweight is known.

United Kingdom by countries and, within England, Government Office Regions; counties/unitary authorities, districts and London boroughs/Northern Ireland council districts

Deaths				Deaths under 1 year				Stillbirth rate	Perinatal mortality rate	Infant mortality rate	Area
Male	Female	Total	SMR	Stillbirths	Under 1 week	Under 4 weeks	Under 1 year				
12,878	13,206	26,084	103	203	130	161	248	6.1	10.0	7.5	**West Midlands (Met County)**
4,851	4,874	9,725	106	93	65	83	127	6.4	10.9	8.8	Birmingham
1,442	1,477	2,919	99	21	14	15	27	5.8	9.6	7.5	Coventry
1,568	1,610	3,178	101	20	7	10	16	6.0	8.1	*4.8*	Dudley
1,609	1,716	3,325	116	23	9	13	25	6.2	8.6	6.8	Sandwell
898	970	1,868	89	13	6	8	9	*6.6*	*9.6*	*4.6*	Solihull
1,225	1,314	2,539	102	17	11	12	18	*5.3*	8.7	*5.6*	Walsall
1,285	1,245	2,530	102	16	18	20	26	*5.5*	11.7	9.0	Wolverhampton
2,616	2,960	5,576	96	23	15	18	22	4.1	6.8	4.0	**Worcestershire**
434	538	972	98	4	1	1	1	*5.0*	*6.2*	*1.2*	Bromsgrove
416	492	908	91	2	2	2	4	*3.4*	*6.7*	*6.7*	Malvern Hills
299	323	622	97	2	1	1	1	*2.1*	*3.1*	*1.0*	Redditch
413	441	854	99	5	5	7	8	*4.3*	*8.6*	*6.9*	Worcester
543	588	1,131	89	6	4	5	6	*5.5*	*9.2*	*5.6*	Wychavon
511	578	1,089	103	4	2	2	2	*4.1*	*6.2*	*2.1*	Wyre Forest
25,264	**28,023**	**53,287**	**92**	**291**	**136**	**176**	**271**	**4.8**	**7.1**	**4.5**	**EAST**
750	811	1,561	110	26	8	10	18	9.0	11.8	*6.3*	**Luton UA**
717	766	1,483	105	15	4	6	13	*7.2*	*9.2*	*6.3*	**Peterborough UA**
939	1,258	2,197	104	14	1	4	6	*7.3*	*7.8*	*3.2*	**Southend-on-Sea UA**
568	592	1,160	99	11	1	2	5	*6.1*	*6.6*	*2.8*	**Thurrock UA**
1,597	1,708	3,305	95	25	12	14	23	5.5	8.2	5.1	**Bedfordshire**
643	716	1,359	94	12	6	7	11	*7.0*	*10.5*	*6.5*	Bedford
469	466	935	89	7	2	2	3	*4.9*	*6.3*	*2.1*	Mid Bedfordshire
485	526	1,011	101	6	4	5	9	*4.4*	*7.3*	*6.6*	South Bedfordshire
2,376	2,471	4,847	89	18	15	16	26	*3.1*	5.6	4.4	**Cambridgeshire**
427	479	906	90	3	2	2	3	*2.8*	*4.7*	*2.8*	Cambridge
336	293	629	81	3	5	6	8	*3.9*	*10.5*	*10.5*	East Cambridgeshire
521	533	1,054	106	6	4	4	6	*7.1*	*11.8*	*7.1*	Fenland
607	604	1,211	89	3	3	3	7	*1.7*	*3.3*	*3.9*	Huntingdonshire
485	562	1,047	80	3	1	1	2	*2.2*	*2.9*	*1.5*	South Cambridgeshire
6,062	6,905	12,967	92	56	26	39	61	4.0	5.8	4.3	**Essex**
683	728	1,411	93	10	6	9	14	*4.8*	*7.7*	*6.8*	Basildon
593	763	1,356	101	8	2	3	5	*5.3*	*6.6*	*3.3*	Braintree
298	383	681	85	4	1	1	2	*5.9*	*7.4*	*3.0*	Brentwood
373	449	822	88	1	1	1	1	*1.2*	*2.4*	*1.2*	Castle Point
638	640	1,278	85	4	2	7	9	*2.5*	*3.7*	*5.6*	Chelmsford
669	739	1,408	91	8	2	3	6	*4.8*	*6.0*	*3.6*	Colchester
554	623	1,177	88	2	2	3	4	*1.5*	*2.9*	*2.9*	Epping Forest
310	332	642	95	6	3	4	5	*5.5*	*8.2*	*4.6*	Harlow
253	318	571	94	3	-	-	1	*4.9*	*4.9*	*1.6*	Maldon
354	369	723	84	-	1	1	3	-	*1.2*	*3.7*	Rochford
1,029	1,178	2,207	94	8	4	5	8	*7.0*	*10.5*	*7.0*	Tendring
308	383	691	97	2	2	2	3	*2.8*	*5.7*	*4.3*	Uttlesford

Table 4.2 - *continued*

Area	Estimated resident population at 30 June 2001 (thousands)	Live births							Conceptions in 2000			
		Male	Female	Total	TFR	% outside marriage		Proportion under 2,500 grams[2]	All ages	Under 18	All ages conception rate per 1,000 women aged 15-44	Under 18 conception rate per 1,000 women aged 15-17
						All	% jointly registered, same address[1]					
Hertfordshire	1,034.9	6,466	6,228	12,694	1.70	31.4	72.2	6.3	15,728	593	72.7	32.5
Broxbourne	87.2	578	522	1,100	1.74	36.5	73.9	6.6	1,439	72	79.3	49.1
Dacorum	137.9	835	835	1,670	1.80	32.8	70.6	6.4	2,089	71	73.3	28.3
East Hertfordshire	129.1	838	769	1,607	1.63	26.8	82.3	4.9	1,919	43	70.8	19.9
Hertsmere	94.5	540	563	1,103	1.62	28.5	73.2	5.7	1,448	59	73.7	37.3
North Hertfordshire	117.1	695	692	1,387	1.76	33.4	70.6	5.6	1,638	57	69.9	27.1
St Albans	129.1	827	846	1,673	1.72	20.8	76.4	5.9	2,011	46	76.0	21.0
Stevenage	79.8	512	482	994	1.72	47.3	67.2	8.3	1,243	85	71.0	53.9
Three Rivers	82.9	492	465	957	1.70	28.2	67.8	7.2	1,212	45	73.0	29.7
Watford	79.7	582	532	1,114	1.68	32.6	70.8	7.2	1,409	53	76.5	40.5
Welwyn Hatfield	97.6	567	522	1,089	1.60	35.0	69.3	6.3	1,320	62	64.3	34.1
Norfolk	797.9	3,756	3,649	7,405	1.56	42.4	72.9	6.8	9,249	496	63.0	37.1
Breckland	121.6	578	543	1,121	1.62	36.6	75.1	6.2	1,373	56	63.1	27.4
Broadland	118.8	555	534	1,089	1.55	30.5	82.8	6.0	1,322	40	61.0	20.7
Great Yarmouth	90.9	483	447	930	1.80	55.5	70.0	7.6	1,156	99	69.7	60.7
King's Lynn and West Norfolk	135.6	648	660	1,308	1.77	45.1	73.9	7.1	1,613	111	67.4	48.4
North Norfolk	98.5	354	335	689	1.48	40.5	72.8	5.4	877	44	56.8	28.2
Norwich	121.7	652	614	1,266	1.27	57.0	65.6	8.5	1,720	109	62.2	55.5
South Norfolk	110.8	486	516	1,002	1.62	29.3	79.6	5.8	1,188	37	60.1	18.9
Suffolk	669.4	3,556	3,400	6,956	1.72	35.7	72.8	6.5	8,250	363	64.7	30.7
Babergh	83.5	352	378	730	1.52	31.1	78.9	5.5	919	42	61.4	27.8
Forest Heath	55.6	395	381	776	1.93	26.5	71.4	5.9	904	23	66.9	22.0
Ipswich	117.2	724	666	1,390	1.69	45.5	68.1	7.1	1,698	101	70.5	44.9
Mid Suffolk	87.0	471	409	880	1.77	31.0	81.7	5.1	985	29	62.0	19.5
St Edmundsbury	98.3	584	528	1,112	1.76	31.4	73.9	6.5	1,262	45	66.5	29.3
Suffolk Coastal	115.2	499	479	978	1.59	28.3	72.2	7.3	1,197	38	58.7	18.6
Waveney	112.5	531	559	1,090	1.77	47.7	71.5	7.1	1,285	85	65.2	43.4
LONDON	**7,188.0**	**53,157**	**51,005**	**104,162**	**1.62**	**34.6**	**53.1**	**8.1**	**153,533**	**6,041**	**87.8**	**50.8**
Inner London	2,771.7	22,989	22,100	45,089	1.63	36.6	47.5	8.7	70,697	2,861	95.3	67.7
Camden	198.4	1,417	1,375	2,792	1.36	31.6	51.3	6.9	4,327	134	80.7	49.2
City of London[3]	7.2	26	27	53	0.80	*30.2*	*87.5*	*7.5*	102	:	63.0	:
Hackney	203.4	2,070	2,025	4,095	2.08	40.5	45.2	9.1	6,140	307	114.1	81.4
Hammersmith and Fulham	165.5	1,256	1,109	2,365	1.38	32.0	46.7	7.0	3,886	126	83.0	63.8
Haringey	216.8	1,954	1,876	3,830	1.80	38.3	47.1	8.6	5,863	273	100.2	73.1
Islington	176.1	1,270	1,230	2,500	1.38	43.4	49.9	7.9	4,256	172	86.1	64.2
Kensington and Chelsea	159.1	1,072	1,066	2,138	1.35	21.7	51.4	7.4	3,303	80	84.7	51.4
Lambeth	266.8	2,218	2,179	4,397	1.65	48.7	44.0	9.6	7,625	358	102.7	89.8
Lewisham	249.5	1,909	1,809	3,718	1.60	48.9	48.3	8.6	5,863	275	90.9	68.6
Newham	244.3	2,463	2,342	4,805	2.19	31.9	41.4	9.7	7,004	312	116.2	55.6
Southwark	245.4	2,032	1,966	3,998	1.67	51.1	45.8	9.8	6,875	314	106.2	83.6
Tower Hamlets	196.6	1,835	1,811	3,646	1.81	21.1	51.1	10.9	5,063	189	98.7	48.3
Wandsworth	260.8	2,193	1,989	4,182	1.47	29.8	55.1	6.9	6,092	202	79.4	70.8
Westminster	181.7	1,274	1,296	2,570	1.39	24.7	53.4	7.7	4,298	119	91.1	72.5

Note: See Appendix A for fuller descriptions and definitions of table headings.

1 Number of live births outside marriage, which are registered by both parents who gave the same address of usual residence, expressed as a percentage of all live births outside marriage.
2 Number of live births under 2,500 grams as a percentage of all live births for which the birthweight is known.
3 To protect confidentiality, teenage conceptions for the City of London have been included with those for Hackney borough.

United Kingdom by countries and, within England, Government Office Regions; counties/unitary authorities, districts and London boroughs/Northern Ireland council districts

Deaths				Deaths under 1 year				Stillbirth rate	Perinatal mortality rate	Infant mortality rate	Area
Male	Female	Total	SMR	Stillbirths	Under 1 week	Under 4 weeks	Under 1 year				
4,386	5,067	9,453	93	66	32	42	60	5.2	7.7	4.7	**Hertfordshire**
361	342	703	90	8	1	3	6	7.2	8.1	5.5	Broxbourne
600	625	1,225	92	5	4	7	8	3.0	5.4	4.8	Dacorum
489	546	1,035	89	6	3	4	5	3.7	5.6	3.1	East Hertfordshire
387	595	982	95	5	2	2	5	4.5	6.3	4.5	Hertsmere
603	720	1,323	106	7	3	3	4	5.0	7.2	2.9	North Hertfordshire
472	573	1,045	83	7	4	7	7	4.2	6.5	4.2	St Albans
320	333	653	98	7	2	3	4	7.0	9.0	4.0	Stevenage
371	418	789	87	7	6	6	6	7.3	13.5	6.3	Three Rivers
341	437	778	109	4	4	4	10	3.6	7.2	9.0	Watford
442	478	920	87	10	3	3	5	9.1	11.8	4.6	Welwyn Hatfield
4,420	4,700	9,120	89	35	20	26	33	4.7	7.4	4.5	**Norfolk**
700	694	1,394	93	5	5	5	6	4.4	8.9	5.4	Breckland
602	706	1,308	88	5	4	4	4	4.6	8.2	3.7	Broadland
517	611	1,128	98	2	3	5	5	2.1	5.4	5.4	Great Yarmouth
833	814	1,647	92	5	2	2	4	3.8	5.3	3.1	King's Lynn and West Norfolk
648	703	1,351	85	2	-	-	2	2.9	2.9	2.9	North Norfolk
567	600	1,167	86	9	6	9	10	7.1	11.8	7.9	Norwich
553	572	1,125	81	7	-	1	2	6.9	6.9	2.0	South Norfolk
3,449	3,745	7,194	90	25	17	17	26	3.6	6.0	3.7	**Suffolk**
415	445	860	87	4	1	1	2	5.4	6.8	2.7	Babergh
261	258	519	96	1	5	5	5	1.3	7.7	6.4	Forest Heath
551	668	1,219	95	7	2	2	5	5.0	6.4	3.6	Ipswich
424	439	863	87	3	2	2	2	3.4	5.7	2.3	Mid Suffolk
467	504	971	92	6	2	2	2	5.4	7.2	1.8	St Edmundsbury
644	712	1,356	87	-	4	4	6	-	4.1	6.1	Suffolk Coastal
687	719	1,406	90	4	1	1	4	3.7	4.6	3.7	Waveney
28,308	**29,858**	**58,166**	**97**	**620**	**314**	**428**	**637**	**5.9**	**8.9**	**6.1**	**LONDON**
10,147	9,684	19,831	104	275	150	198	305	6.1	9.4	6.8	**Inner London**
769	737	1,506	104	14	7	8	17	5.0	7.5	6.1	Camden
19	14	33	49	-	-	-	-	-	-	-	City of London
672	606	1,278	100	28	18	24	32	6.8	11.2	7.8	Hackney
548	533	1,081	92	10	5	7	13	4.2	6.3	5.5	Hammersmith and Fulham
712	763	1,475	104	31	13	18	28	8.0	11.4	7.3	Haringey
723	603	1,326	114	20	10	12	15	7.9	11.9	6.0	Islington
492	495	987	73	12	4	6	13	5.6	7.4	6.1	Kensington and Chelsea
971	838	1,809	109	32	14	20	29	7.2	10.4	6.6	Lambeth
1,038	1,091	2,129	114	23	13	21	30	6.1	9.6	8.1	Lewisham
896	773	1,669	114	31	18	23	33	6.4	10.1	6.9	Newham
887	871	1,758	105	26	22	24	41	6.5	11.9	10.3	Southwark
754	605	1,359	117	17	8	11	19	4.6	6.8	5.2	Tower Hamlets
973	1,098	2,071	111	16	9	12	17	3.8	6.0	4.1	Wandsworth
693	657	1,350	90	15	9	12	18	5.8	9.3	7.0	Westminster

Table 4.2 - *continued*

Area	Estimated resident population at 30 June 2001 (thousands)	Live births: Male	Live births: Female	Live births: Total	Live births: TFR	% outside marriage: All	% outside marriage: % jointly registered, same address[1]	Proportion under 2,500 grams[2]	Conceptions in 2000: All ages	Conceptions in 2000: Under 18	All ages conception rate per 1,000 women aged 15-44	Under 18 conception rate per 1,000 women aged 15-17
Outer London	4,416.4	30,168	28,905	59,073	1.64	33.1	57.7	7.7	82,836	3,180	82.3	41.5
Barking and Dagenham	164.3	1,211	1,196	2,407	1.87	49.6	57.2	7.2	3,237	216	87.6	66.1
Barnet	315.3	2,123	1,940	4,063	1.54	26.0	56.0	7.6	5,641	154	77.8	28.7
Bexley	218.8	1,324	1,302	2,626	1.73	39.8	67.6	7.1	3,329	139	73.4	35.2
Brent	263.8	1,941	1,976	3,917	1.61	30.7	45.1	9.1	6,558	259	99.4	53.5
Bromley	296.2	1,731	1,683	3,414	1.58	34.6	66.3	7.1	4,526	160	74.8	35.1
Croydon	331.5	2,216	2,185	4,401	1.67	41.5	50.5	8.8	6,442	347	84.0	56.0
Ealing	301.6	2,255	2,137	4,392	1.62	25.7	52.3	8.6	6,322	203	85.3	40.7
Enfield	274.3	1,921	1,826	3,747	1.70	34.2	54.5	7.1	5,474	211	87.2	42.7
Greenwich	215.2	1,608	1,602	3,210	1.74	47.2	51.8	8.4	4,478	222	87.6	57.2
Harrow	208.0	1,351	1,230	2,581	1.58	20.6	59.5	8.2	3,585	84	77.7	21.3
Havering	224.7	1,275	1,107	2,382	1.65	40.4	69.2	6.4	3,170	156	71.3	38.0
Hillingdon	243.1	1,662	1,582	3,244	1.67	34.5	60.3	7.7	4,503	181	81.3	41.7
Hounslow	212.7	1,639	1,495	3,134	1.67	29.7	54.2	7.2	4,343	160	84.3	43.5
Kingston upon Thames	147.6	895	892	1,787	1.44	27.0	71.2	5.9	2,361	75	69.8	32.4
Merton	188.3	1,365	1,299	2,664	1.59	30.4	63.0	6.9	3,595	125	80.2	46.5
Redbridge	239.3	1,633	1,477	3,110	1.69	24.7	56.0	8.0	4,380	113	83.3	25.3
Richmond upon Thames	172.8	1,198	1,196	2,394	1.54	21.3	70.4	6.6	3,074	52	79.1	23.3
Sutton	180.2	1,057	1,033	2,090	1.50	35.1	67.0	7.0	2,759	109	69.5	35.6
Waltham Forest	218.6	1,763	1,747	3,510	1.81	36.4	53.9	8.8	5,059	214	94.6	56.6
SOUTH EAST	**8,006.9**	**45,457**	**43,053**	**88,510**	**1.62**	**34.8**	**70.7**	**6.9**	**111,290**	**5,085**	**68.5**	**36.1**
Bracknell Forest UA	109.6	691	672	1,363	1.59	30.6	74.6	7.6	1,720	68	68.2	35.5
Brighton and Hove UA	248.1	1,428	1,403	2,831	1.34	46.2	70.8	7.1	4,189	188	72.5	49.6
Isle of Wight UA	132.9	591	518	1,109	1.64	46.8	66.3	6.2	1,395	101	63.6	43.5
Medway UA	249.7	1,584	1,444	3,028	1.74	45.3	68.5	7.1	3,866	215	72.6	42.4
Milton Keynes UA	207.6	1,442	1,388	2,830	1.83	41.8	70.6	8.6	3,629	214	77.6	53.0
Portsmouth UA	186.9	1,127	1,026	2,153	1.46	45.1	64.4	9.1	2,771	130	66.1	39.1
Reading UA	143.2	1,013	953	1,966	1.59	38.3	59.0	8.3	2,645	149	77.5	61.0
Slough UA	119.1	952	912	1,864	1.83	31.9	57.6	9.5	2,524	103	87.7	46.8
Southampton UA	217.6	1,272	1,162	2,434	1.39	48.0	65.0	7.8	3,064	201	61.7	57.2
West Berkshire UA	144.5	822	746	1,568	1.58	31.6	76.8	6.8	1,937	83	63.9	28.4
Windsor and Maidenhead UA	133.5	791	817	1,608	1.73	26.2	71.7	7.5	2,026	62	74.6	27.8
Wokingham UA	150.4	830	821	1,651	1.52	20.8	79.7	5.3	2,040	77	64.9	27.2
Buckinghamshire	479.1	2,827	2,636	5,463	1.67	25.5	72.0	6.7	6,686	239	68.2	27.2
Aylesbury Vale	165.9	1,001	906	1,907	1.61	25.8	73.4	6.8	2,356	78	67.1	25.5
Chiltern	89.2	473	469	942	1.74	23.6	73.9	5.2	1,122	35	66.9	22.9
South Bucks	61.9	330	294	624	1.62	25.3	74.1	5.1	793	28	68.6	26.5
Wycombe	162.0	1,023	967	1,990	1.71	26.3	69.4	7.9	2,415	98	69.9	31.3
East Sussex	493.1	2,353	2,211	4,564	1.71	39.5	73.3	6.2	5,714	324	66.4	38.7
Eastbourne	89.8	485	407	892	1.64	46.3	73.6	6.6	1,144	75	69.4	53.1
Hastings	85.4	449	489	938	1.77	53.0	65.6	7.0	1,180	103	71.6	68.9
Lewes	92.3	439	410	849	1.75	37.9	75.2	6.0	1,047	48	66.2	30.3
Rother	85.5	307	301	608	1.62	37.3	71.4	5.3	797	36	61.7	26.9
Wealden	140.2	673	604	1,277	1.73	27.1	83.5	5.7	1,546	62	63.6	24.4
Hampshire	1,240.8	6,639	6,261	12,900	1.63	32.7	71.5	6.5	15,716	690	64.1	31.1
Basingstoke and Deane	152.6	903	869	1,772	1.65	35.5	72.7	6.6	2,175	83	67.0	30.4
East Hampshire	109.4	574	559	1,133	1.78	27.3	72.8	6.3	1,307	53	62.8	26.0
Eastleigh	116.3	644	548	1,192	1.54	30.8	71.7	5.3	1,461	57	61.1	26.2
Fareham	108.1	538	542	1,080	1.68	29.4	74.1	7.1	1,308	55	64.2	28.7
Gosport	76.4	439	399	838	1.61	44.2	68.1	6.8	1,114	66	68.1	45.6

Note: See Appendix A for fuller descriptions and definitions of table headings.

1 Number of live births outside marriage, which are registered by both parents who gave the same address of usual residence, expressed as a percentage of all live births outside marriage.

2 Number of live births under 2,500 grams as a percentage of all live births for which the birthweight is known.

United Kingdom by countries and, within England, Government Office Regions; counties/unitary authorities, districts and London boroughs/Northern Ireland council districts

Deaths				Deaths under 1 year				Stillbirth rate	Perinatal mortality rate	Infant mortality rate	Area
Male	Female	Total	SMR	Stillbirths	Under 1 week	Under 4 weeks	Under 1 year				
18,161	20,174	38,335	94	345	164	230	332	5.8	8.6	5.6	**Outer London**
845	860	1,705	110	17	4	5	8	*7.0*	8.7	*3.3*	Barking and Dagenham
1,267	1,548	2,815	89	30	7	11	14	7.3	9.0	*3.4*	Barnet
995	1,059	2,054	93	14	8	12	14	*5.3*	8.3	*5.3*	Bexley
926	862	1,788	92	23	20	24	34	5.8	10.9	8.7	Brent
1,352	1,570	2,922	90	15	7	8	11	*4.4*	6.4	*3.2*	Bromley
1,225	1,420	2,645	92	26	9	12	19	5.9	7.9	*4.3*	Croydon
1,127	1,116	2,243	96	28	13	18	26	6.3	9.3	5.9	Ealing
1,091	1,293	2,384	93	25	12	16	20	6.6	9.8	5.3	Enfield
921	1,053	1,974	106	22	9	13	21	6.8	9.6	6.5	Greenwich
770	850	1,620	79	14	5	7	10	*5.4*	*7.3*	*3.9*	Harrow
1,113	1,222	2,335	96	6	3	5	7	*2.5*	*3.8*	*2.9*	Havering
989	1,089	2,078	94	16	10	13	21	*4.9*	8.0	6.5	Hillingdon
833	855	1,688	104	15	14	20	29	*4.8*	9.2	9.3	Hounslow
610	733	1,343	96	11	2	3	7	*6.1*	*7.2*	*3.9*	Kingston upon Thames
722	785	1,507	92	15	2	6	11	*5.6*	*6.3*	*4.1*	Merton
997	1,094	2,091	93	20	10	16	23	6.4	9.6	7.4	Redbridge
662	774	1,436	85	14	7	11	14	*5.8*	8.7	*5.8*	Richmond upon Thames
779	995	1,774	100	8	3	5	9	*3.8*	*5.2*	*4.3*	Sutton
937	996	1,933	110	26	19	25	34	7.4	12.7	9.7	Waltham Forest
36,882	**42,566**	**79,448**	**91**	**409**	**201**	**261**	**376**	**4.6**	**6.9**	**4.2**	**SOUTH EAST**
369	398	767	96	8	3	6	6	*5.8*	*8.0*	*4.4*	**Bracknell Forest UA**
1,328	1,431	2,759	98	16	5	7	8	*5.6*	7.4	*2.8*	**Brighton and Hove UA**
820	939	1,759	90	6	-	1	3	*5.4*	*5.4*	*2.7*	**Isle of Wight UA**
1,045	1,109	2,154	105	15	4	5	9	*4.9*	*6.2*	*3.0*	**Medway UA**
696	784	1,480	101	14	15	17	23	*4.9*	10.2	8.1	**Milton Keynes UA**
903	1,058	1,961	101	18	7	13	19	*8.3*	11.5	*8.8*	**Portsmouth UA**
555	599	1,154	94	10	14	14	17	*5.1*	12.1	*8.6*	**Reading UA**
500	450	950	106	10	5	7	10	*5.3*	*8.0*	*5.4*	**Slough UA**
906	981	1,887	90	8	6	10	14	*3.3*	*5.7*	*5.8*	**Southampton UA**
547	593	1,140	89	4	7	7	12	*2.5*	*7.0*	*7.7*	**West Berkshire UA**
612	687	1,299	96	12	5	6	9	*7.4*	*10.5*	*5.6*	**Windsor and Maidenhead UA**
523	509	1,032	86	4	1	1	2	*2.4*	*3.0*	*1.2*	**Wokingham UA**
1,981	2,158	4,139	90	30	23	27	32	5.5	9.6	5.9	**Buckinghamshire**
603	723	1,326	93	13	6	6	7	*6.8*	*9.9*	*3.7*	Aylesbury Vale
416	418	834	85	4	1	1	4	*4.2*	*5.3*	*4.2*	Chiltern
302	337	639	90	4	2	3	3	*6.4*	*9.6*	*4.8*	South Bucks
660	680	1,340	89	9	14	17	18	*4.5*	11.5	*9.0*	Wycombe
3,069	3,764	6,833	90	15	10	15	24	*3.3*	5.5	5.3	**East Sussex**
669	751	1,420	92	2	3	3	5	*2.2*	*5.6*	*5.6*	Eastbourne
483	664	1,147	109	1	3	4	8	*1.1*	*4.3*	*8.5*	Hastings
518	621	1,139	82	4	-	-	2	*4.7*	*4.7*	*2.4*	Lewes
649	833	1,482	90	4	-	-	-	*6.5*	*6.5*	-	Rother
750	895	1,645	83	4	4	8	9	*3.1*	*6.2*	*7.0*	Wealden
5,364	6,149	11,513	87	65	20	32	52	5.0	6.6	4.0	**Hampshire**
533	586	1,119	89	7	4	4	8	*3.9*	*6.2*	*4.5*	Basingstoke and Deane
459	609	1,068	91	4	-	-	1	*3.5*	*3.5*	*0.9*	East Hampshire
463	526	989	88	6	2	3	3	*5.0*	*6.7*	*2.5*	Eastleigh
477	573	1,050	88	5	2	5	5	*4.6*	*6.5*	*4.6*	Fareham
390	413	803	101	9	1	2	2	*10.6*	*11.8*	*2.4*	Gosport

Table 4.2 - *continued*

Area	Estimated resident population at 30 June 2001 (thousands)	Live births							Conceptions in 2000			
		Male	Female	Total	TFR	% outside marriage		Proportion under 2,500 grams[2]	All ages	Under 18	All ages conception rate per 1,000 women aged 15-44	Under 18 conception rate per 1,000 women aged 15-17
						All	% jointly registered, same address[1]					
Hart	83.6	495	402	897	1.60	21.9	77.6	5.9	1,067	25	65.4	17.1
Havant	116.9	594	596	1,190	1.78	46.1	64.4	7.9	1,473	112	66.2	49.1
New Forest	169.5	753	701	1,454	1.62	34.0	72.9	7.7	1,764	92	60.3	33.6
Rushmoor	90.9	615	562	1,177	1.63	34.2	71.4	5.7	1,465	59	71.3	38.5
Test Valley	109.9	560	557	1,117	1.59	28.8	70.5	6.1	1,318	47	61.5	24.9
Winchester	107.3	524	526	1,050	1.55	25.4	76.8	5.9	1,264	41	59.2	20.8
Kent	1,331.1	7,523	7,121	14,644	1.74	40.9	69.7	6.9	18,228	1,055	70.3	43.3
Ashford	103.0	651	599	1,250	1.90	40.6	74.2	5.4	1,533	93	76.7	50.9
Canterbury	135.4	634	634	1,268	1.45	40.5	69.2	6.5	1,641	112	59.6	43.4
Dartford	86.0	556	517	1,073	1.68	42.7	72.7	6.1	1,345	72	72.4	49.3
Dover	104.6	565	513	1,078	1.81	44.9	68.8	6.7	1,306	87	66.7	45.4
Gravesham	95.8	577	521	1,098	1.77	42.6	69.0	8.8	1,403	92	73.4	48.8
Maidstone	139.1	781	776	1,557	1.67	32.8	70.6	6.0	1,911	92	69.4	36.8
Sevenoaks	109.2	601	594	1,195	1.81	28.6	72.8	5.7	1,504	50	73.5	25.1
Shepway	96.4	513	471	984	1.80	49.9	63.1	6.6	1,217	80	68.5	48.9
Swale	123.1	734	696	1,430	1.85	50.0	67.1	8.0	1,753	117	73.1	51.6
Thanet	126.8	656	629	1,285	1.80	55.1	65.4	8.9	1,667	163	73.2	70.8
Tonbridge and Malling	107.8	654	621	1,275	1.87	32.3	70.6	7.5	1,491	48	70.6	25.3
Tunbridge Wells	104.1	601	550	1,151	1.70	33.1	78.5	6.7	1,457	49	70.3	23.3
Oxfordshire	605.9	3,494	3,525	7,019	1.54	29.8	71.3	6.8	8,784	320	66.5	30.0
Cherwell	132.0	850	829	1,679	1.74	32.2	74.5	8.8	2,034	79	70.7	33.0
Oxford	134.1	767	738	1,505	1.32	36.0	58.7	7.4	2,146	89	59.4	36.3
South Oxfordshire	128.3	740	809	1,549	1.73	25.6	78.8	5.8	1,798	45	69.9	20.9
Vale of White Horse	115.8	627	609	1,236	1.64	26.5	72.8	5.0	1,530	62	68.0	30.3
West Oxfordshire	95.7	510	540	1,050	1.68	27.5	77.2	6.5	1,276	45	67.6	28.0
Surrey	1,059.5	6,094	5,668	11,762	1.57	26.3	75.8	6.2	14,816	455	69.1	25.9
Elmbridge	122.0	766	702	1,468	1.65	22.9	79.8	5.7	1,872	56	76.6	28.5
Epsom and Ewell	67.1	365	360	725	1.56	25.8	71.1	6.4	948	33	71.1	29.6
Guildford	129.8	727	657	1,384	1.43	26.0	69.7	6.4	1,716	41	61.8	20.4
Mole Valley	80.3	417	382	799	1.63	25.2	75.6	5.4	956	22	66.0	16.9
Reigate and Banstead	126.7	727	715	1,442	1.62	26.1	76.9	6.0	1,783	54	71.0	26.8
Runnymede	78.0	430	439	869	1.48	33.3	75.8	7.5	1,140	52	65.9	42.3
Spelthorne	90.4	553	479	1,032	1.60	36.2	78.1	6.2	1,340	46	71.8	35.2
Surrey Heath	80.3	460	414	874	1.53	22.1	79.3	5.9	1,108	32	66.3	22.5
Tandridge	79.3	422	433	855	1.70	29.2	76.8	7.3	1,062	42	68.3	27.6
Waverley	115.6	653	556	1,209	1.63	22.8	73.2	5.1	1,480	35	67.1	17.0
Woking	89.9	574	531	1,105	1.64	22.6	76.4	7.0	1,411	42	73.8	26.6
West Sussex	754.3	3,984	3,769	7,753	1.68	33.9	72.6	6.9	9,540	411	67.4	33.1
Adur	59.7	297	280	577	1.71	37.1	75.2	8.5	696	38	66.4	41.9
Arun	141.0	627	677	1,304	1.75	38.3	74.6	5.9	1,572	85	66.5	42.4
Chichester	106.5	499	457	956	1.62	32.8	72.9	5.2	1,142	44	61.0	24.6
Crawley	99.7	660	584	1,244	1.62	35.8	68.5	8.5	1,688	92	74.9	53.1
Horsham	122.3	630	625	1,255	1.64	28.8	74.5	7.4	1,528	41	64.9	18.9
Mid Sussex	127.4	700	666	1,366	1.67	27.9	74.0	6.7	1,635	33	66.4	14.3
Worthing	97.6	571	480	1,051	1.74	39.2	70.1	6.4	1,279	78	70.5	52.3

Note: See Appendix A for fuller descriptions and definitions of table headings.

1 Number of live births outside marriage, which are registered by both parents who gave the same address of usual residence, expressed as a percentage of all live births outside marriage.

2 Number of live births under 2,500 grams as a percentage of all live births for which the birthweight is known.

United Kingdom by countries and, within England, Government Office Regions; counties/unitary authorities, districts and London boroughs/Northern Ireland council districts

Deaths				Deaths under 1 year				Stillbirth rate	Perinatal mortality rate	Infant mortality rate	Area
Male	Female	Total	SMR	Stillbirths	Under 1 week	Under 4 weeks	Under 1 year				
274	292	566	79	2	3	3	5	*2.2*	*5.6*	*5.6*	Hart
609	668	1,277	94	9	2	3	6	*7.5*	*9.2*	*5.0*	Havant
867	1,014	1,881	75	10	2	2	4	*6.8*	*8.2*	*2.8*	New Forest
331	373	704	99	4	-	-	1	*3.4*	*3.4*	*0.8*	Rushmoor
483	522	1,005	91	4	4	7	11	*3.6*	*7.1*	*9.8*	Test Valley
478	573	1,051	85	5	-	3	6	*4.7*	*4.7*	*5.7*	Winchester
6,485	7,572	14,057	94	70	32	37	52	4.8	6.9	3.6	**Kent**
464	457	921	85	2	3	3	4	*1.6*	*4.0*	*3.2*	Ashford
712	922	1,634	93	5	4	4	6	*3.9*	*7.1*	*4.7*	Canterbury
376	459	835	105	7	1	1	2	*6.5*	*7.4*	*1.9*	Dartford
596	688	1,284	101	6	4	4	7	*5.5*	*9.2*	*6.5*	Dover
411	441	852	93	4	2	4	5	*3.6*	*5.4*	*4.6*	Gravesham
645	676	1,321	93	5	5	5	7	*3.2*	*6.4*	*4.5*	Maidstone
463	534	997	82	8	4	4	4	*6.7*	*10.0*	*3.3*	Sevenoaks
563	701	1,264	98	2	3	3	4	*2.0*	*5.1*	*4.1*	Shepway
544	667	1,211	102	7	2	2	3	*4.9*	*6.3*	*2.1*	Swale
852	973	1,825	101	8	1	3	6	*6.2*	*7.0*	*4.7*	Thanet
450	465	915	89	8	1	2	2	*6.2*	*7.0*	*1.6*	Tonbridge and Malling
409	589	998	88	8	2	2	2	*6.9*	*8.6*	*1.7*	Tunbridge Wells
2,458	2,638	5,096	87	34	15	17	28	4.8	6.9	4.0	**Oxfordshire**
523	591	1,114	94	9	2	3	4	*5.3*	*6.5*	*2.4*	Cherwell
490	510	1,000	84	11	4	5	12	*7.3*	*9.9*	*8.0*	Oxford
567	568	1,135	86	6	5	5	6	*3.9*	*7.1*	*3.9*	South Oxfordshire
459	478	937	80	2	2	2	3	*1.6*	*3.2*	*2.4*	Vale of White Horse
419	491	910	89	6	2	2	3	*5.7*	*7.6*	*2.9*	West Oxfordshire
4,609	5,495	10,104	87	34	18	23	35	2.9	4.4	3.0	**Surrey**
519	663	1,182	86	5	-	-	1	*3.4*	*3.4*	*0.7*	Elmbridge
353	311	664	86	1	-	-	1	*1.4*	*1.4*	*1.4*	Epsom and Ewell
479	541	1,020	78	5	1	1	2	*3.6*	*4.3*	*1.4*	Guildford
390	469	859	84	2	3	4	4	*2.5*	*6.2*	*5.0*	Mole Valley
624	776	1,400	100	5	4	4	5	*3.5*	*6.2*	*3.5*	Reigate and Banstead
362	413	775	93	2	1	4	5	*2.3*	*3.4*	*5.8*	Runnymede
381	433	814	86	4	1	2	4	*3.9*	*4.8*	*3.9*	Spelthorne
318	347	665	93	-	2	2	2	-	*2.3*	*2.3*	Surrey Heath
345	436	781	87	4	1	1	2	*4.7*	*5.8*	*2.3*	Tandridge
491	692	1,183	83	3	3	3	4	*2.5*	*5.0*	*3.3*	Waverley
347	414	761	87	3	2	2	5	*2.7*	*4.5*	*4.5*	Woking
4,112	5,252	9,364	91	36	11	16	21	4.6	6.0	2.7	**West Sussex**
309	412	721	84	4	1	1	1	*6.9*	*8.6*	*1.7*	Adur
999	1,306	2,305	93	5	2	2	2	*3.8*	*5.3*	*1.5*	Arun
647	785	1,432	89	3	-	1	1	*3.1*	*3.1*	*1.0*	Chichester
367	419	786	88	3	3	4	5	*2.4*	*4.8*	*4.0*	Crawley
559	589	1,148	83	7	-	1	2	*5.5*	*5.5*	*1.6*	Horsham
557	776	1,333	93	5	3	3	5	*3.6*	*5.8*	*3.7*	Mid Sussex
674	965	1,639	100	9	2	4	5	*8.5*	*10.4*	*4.8*	Worthing

Table 4.2 - *continued*

Area	Estimated resident population at 30 June 2001 (thousands)	Live births							Conceptions in 2000			
		Male	Female	Total	TFR	% outside marriage		Proportion under 2,500 grams[2]	All ages	Under 18	All ages conception rate per 1,000 women aged 15-44	Under 18 conception rate per 1,000 women aged 15-17
						All	% jointly registered, same address[1]					
SOUTH WEST	**4,934.2**	**24,938**	**23,809**	**48,747**	**1.59**	**39.3**	**71.1**	**6.7**	**60,231**	**3,092**	**64.0**	**36.3**
Bath and North East Somerset UA	169.2	882	780	1,662	1.45	35.3	68.1	5.6	2,041	78	59.8	27.9
Bournemouth UA	163.6	809	731	1,540	1.32	40.6	67.9	7.2	2,023	107	60.9	44.4
Bristol, City of UA	380.8	2,370	2,325	4,695	1.51	45.7	64.4	7.3	6,012	349	67.8	52.8
North Somerset UA	188.8	916	952	1,868	1.74	38.2	71.6	6.9	2,213	108	65.4	33.7
Plymouth UA	241.0	1,320	1,227	2,547	1.54	48.7	66.6	6.6	3,138	217	62.0	49.2
Poole UA	138.4	670	669	1,339	1.55	37.6	70.4	6.5	1,649	75	61.9	31.0
South Gloucestershire UA	246.0	1,399	1,285	2,684	1.56	32.0	77.0	6.0	3,251	122	64.2	29.7
Swindon UA	180.2	1,135	1,104	2,239	1.73	42.5	72.8	9.0	2,786	160	71.2	51.6
Torbay UA	130.0	543	526	1,069	1.53	47.2	69.1	7.8	1,437	105	64.5	49.1
Cornwall and the Isles of Scilly	502.1	2,298	2,179	4,477	1.65	44.1	71.9	6.5	5,505	309	62.7	35.4
Caradon	79.7	364	342	706	1.73	41.4	68.5	6.5	874	59	61.9	39.0
Carrick	88.1	401	367	768	1.52	44.5	74.9	5.7	940	44	60.5	29.3
Kerrier	92.7	472	408	880	1.71	46.0	69.6	6.9	1,074	58	64.3	37.3
North Cornwall	80.7	360	352	712	1.74	44.5	75.1	5.6	848	40	62.1	27.8
Penwith	63.0	227	250	477	1.49	52.4	68.4	5.5	616	44	59.2	40.2
Restormel	95.8	465	455	920	1.70	39.7	74.2	8.0	1,128	64	66.2	39.5
Isles of Scilly	2.1	9	5	14	*0.94*	*14.3*	*50.0*	-	25	-	65.1	-
Devon	705.6	3,191	3,036	6,227	1.57	39.2	72.7	5.9	7,617	364	61.0	30.4
East Devon	125.7	468	475	943	1.54	38.6	73.1	5.1	1,218	60	62.0	31.2
Exeter	111.2	550	559	1,109	1.30	45.9	66.2	6.2	1,389	72	55.5	38.1
Mid Devon	69.9	351	344	695	1.83	36.4	78.7	5.9	810	41	64.6	31.4
North Devon	87.7	438	378	816	1.76	42.2	71.5	7.0	952	39	62.8	26.9
South Hams	81.9	362	297	659	1.62	34.9	73.5	5.3	833	36	60.5	25.2
Teignbridge	121.2	535	546	1,081	1.67	36.5	73.4	4.9	1,355	65	64.7	31.8
Torridge	59.2	255	241	496	1.64	40.3	73.5	7.3	568	29	57.4	27.0
West Devon	48.9	232	196	428	1.79	33.9	82.8	6.8	492	22	61.6	26.0
Dorset	391.5	1,666	1,625	3,291	1.67	35.6	72.6	6.6	3,986	180	61.2	27.2
Christchurch	44.9	190	188	378	1.74	32.3	69.7	*4.8*	438	22	63.9	34.4
East Dorset	83.9	329	311	640	1.61	26.7	75.4	6.3	753	29	56.5	21.6
North Dorset	62.0	260	248	508	1.56	31.7	75.2	7.9	626	30	58.4	26.8
Purbeck	44.4	203	188	391	1.73	37.6	74.8	6.4	480	18	62.9	*24.2*
West Dorset	92.5	386	403	789	1.74	34.2	72.6	6.9	921	37	60.7	21.9
Weymouth and Portland	63.8	298	287	585	1.63	51.1	69.9	6.8	768	44	67.2	40.0
Gloucestershire	565.0	2,964	2,818	5,782	1.61	38.4	71.5	6.8	7,211	353	65.0	35.3
Cheltenham	110.0	560	548	1,108	1.39	42.2	66.0	7.0	1,376	76	58.4	35.6
Cotswold	80.4	349	374	723	1.59	28.4	79.5	5.8	915	26	62.8	20.1
Forest of Dean	80.0	398	331	729	1.56	36.1	79.5	6.7	939	47	63.1	31.5
Gloucester	109.9	689	630	1,319	1.75	44.4	63.9	8.3	1,729	112	74.0	59.9
Stroud	108.1	566	532	1,098	1.77	38.4	77.7	5.8	1,314	55	65.9	28.4
Tewkesbury	76.5	402	403	805	1.71	34.5	73.7	6.5	938	37	64.5	28.8

Notes: See Appendix A for fuller descriptions and definitions of table headings.

1 Number of live births outside marriage, which are registered by both parents who gave the same address of usual residence, expressed as a percentage of all live births outside marriage.
2 Number of live births under 2,500 grams as a percentage of all live births for which the birthweight is known.

United Kingdom by countries and, within England, Government Office Regions; counties/unitary authorities, districts and London boroughs/Northern Ireland council districts

Deaths				Deaths under 1 year				Stillbirth rate	Perinatal mortality rate	Infant mortality rate	Area
Male	Female	Total	SMR	Stillbirths	Under 1 week	Under 4 weeks	Under 1 year				
25,620	**28,667**	**54,287**	**90**	**211**	**142**	**180**	**263**	**4.3**	**7.2**	**5.4**	**SOUTH WEST**
806	831	1,637	83	6	1	3	5	*3.6*	*4.2*	*3.0*	**Bath and North East Somerset UA**
1,017	1,268	2,285	96	4	6	8	11	*2.6*	*6.5*	*7.1*	**Bournemouth UA**
1,844	1,975	3,819	102	19	16	20	33	*4.0*	7.4	7.0	**Bristol, City of UA**
1,011	1,245	2,256	92	6	7	9	14	*3.2*	*6.9*	*7.5*	**North Somerset UA**
1,187	1,293	2,480	99	12	7	8	11	*4.7*	*7.4*	*4.3*	**Plymouth UA**
706	813	1,519	82	5	5	5	7	*3.7*	*7.4*	*5.2*	**Poole UA**
871	996	1,867	83	12	9	11	16	*4.5*	7.8	*6.0*	**South Gloucestershire UA**
798	789	1,587	101	17	9	11	15	*7.5*	11.5	*6.7*	**Swindon UA**
838	1,075	1,913	94	6	3	3	3	*5.6*	*8.4*	*2.8*	**Torbay UA**
2,822	3,109	5,931	90	14	12	14	23	*3.1*	5.8	5.1	**Cornwall and the Isles of Scilly**
414	490	904	90	-	-	-	2	-	-	*2.8*	Caradon
495	571	1,066	86	2	3	4	7	*2.6*	*6.5*	*9.1*	Carrick
503	551	1,054	93	3	2	3	3	*3.4*	*5.7*	*3.4*	Kerrier
469	489	958	89	3	2	2	5	*4.2*	*7.0*	*7.0*	North Cornwall
388	437	825	93	3	1	1	1	*6.3*	*8.3*	*2.1*	Penwith
543	560	1,103	92	3	4	4	5	*3.3*	*7.6*	*5.4*	Restormel
10	11	21	79	-	-	-	-	-	-	-	Isles of Scilly
4,007	4,535	8,542	88	25	15	21	33	4.0	6.4	5.3	**Devon**
924	995	1,919	86	4	1	3	3	*4.2*	*5.3*	*3.2*	East Devon
488	592	1,080	90	4	5	5	7	*3.6*	*8.1*	*6.3*	Exeter
361	365	726	85	6	3	3	6	*8.6*	*12.8*	*8.6*	Mid Devon
510	568	1,078	94	4	1	3	6	*4.9*	*6.1*	*7.4*	North Devon
411	497	908	80	3	-	-	2	*4.5*	*4.5*	*3.0*	South Hams
697	799	1,496	86	4	3	4	5	*3.7*	*6.5*	*4.6*	Teignbridge
344	392	736	94	-	2	3	4	-	*4.0*	*8.1*	Torridge
272	327	599	93	-	-	-	-	-	-	-	West Devon
2,326	2,517	4,843	82	16	8	9	13	*4.8*	7.3	*4.0*	**Dorset**
334	344	678	79	1	-	-	-	*2.6*	*2.6*	-	Christchurch
488	525	1,013	75	2	2	2	3	*3.1*	*6.2*	*4.7*	East Dorset
314	331	645	80	6	-	-	-	*11.7*	*11.7*	-	North Dorset
228	256	484	78	1	-	-	-	*2.6*	*2.6*	-	Purbeck
588	652	1,240	85	2	6	7	9	*2.5*	*10.1*	*11.4*	West Dorset
374	409	783	95	4	-	-	1	*6.8*	*6.8*	*1.7*	Weymouth and Portland
2,766	3,063	5,829	90	31	14	21	27	5.3	7.7	4.7	**Gloucestershire**
543	603	1,146	89	8	1	5	8	*7.2*	*8.1*	*7.2*	Cheltenham
441	461	902	86	2	3	3	4	*2.8*	*6.9*	*5.5*	Cotswold
419	425	844	95	3	1	2	2	*4.1*	*5.5*	*2.7*	Forest of Dean
492	507	999	95	11	4	5	6	*8.3*	*11.3*	*4.5*	Gloucester
517	656	1,173	92	6	1	2	3	*5.4*	*6.3*	*2.7*	Stroud
354	411	765	86	1	4	4	4	*1.2*	*6.2*	*5.0*	Tewkesbury

Table 4.2 - *continued*

Area	Estimated resident population at 30 June 2001 (thousands)	Live births: Male	Female	Total	TFR	% outside marriage: All	% jointly registered, same address[1]	Proportion under 2,500 grams[2]	Conceptions in 2000[5]: All ages	Under 18	All ages conception rate per 1,000 women aged 15-44	Under 18 conception rate per 1,000 women aged 15-17
Somerset	498.7	2,415	2,317	4,732	1.68	37.9	73.0	6.5	5,753	333	63.8	36.6
Mendip	104.0	522	470	992	1.63	37.2	79.7	5.3	1,192	68	61.4	32.8
Sedgemoor	106.0	502	481	983	1.69	41.2	70.9	7.0	1,230	89	64.4	47.1
South Somerset	151.1	748	736	1,484	1.78	34.8	70.2	6.9	1,761	87	65.3	33.2
Taunton Deane	102.6	510	506	1,016	1.60	38.2	72.2	6.4	1,255	73	65.0	37.1
West Somerset	35.1	133	124	257	1.65	44.0	75.2	*6.6*	315	16	58.5	*29.3*
Wiltshire	433.5	2,360	2,235	4,595	1.70	31.5	75.2	6.3	5,609	232	67.1	30.9
Kennet	74.9	474	411	885	1.98	30.1	74.8	6.3	1,066	35	73.7	25.9
North Wiltshire	125.4	713	652	1,365	1.65	27.8	76.8	6.2	1,649	59	65.3	27.7
Salisbury	114.7	541	563	1,104	1.56	28.9	76.8	5.1	1,401	64	64.2	32.5
West Wiltshire	118.5	632	609	1,241	1.73	38.8	73.0	7.6	1,493	74	67.9	36.1
WALES	**2,903.2**	**15,746**	**14,870**	**30,616**	**1.66**	**48.3**	**60.8**	**7.5**	**37,892**	**2,649**	**66.3**	**48.4**
Blaenau Gwent	70.0	352	368	720	1.73	63.5	55.4	9.4	899	90	65.7	68.4
Bridgend	128.7	709	735	1,444	1.84	48.4	59.1	6.4	1,756	125	69.3	52.7
Caerphilly	169.6	1,020	954	1,974	1.81	52.9	53.4	8.6	2,376	208	69.5	61.7
Cardiff	305.2	1,883	1,706	3,589	1.46	45.0	53.5	7.7	4,574	266	63.5	46.8
Carmarthenshire	173.7	887	781	1,668	1.68	48.4	67.1	7.4	2,033	124	63.8	37.7
Ceredigion	75.3	303	265	568	1.34	42.4	73.0	5.0	759	50	50.3	36.8
Conwy	109.8	528	540	1,068	1.86	50.2	69.0	7.5	1,377	96	73.4	51.7
Denbighshire	93.1	477	451	928	1.75	48.9	64.5	6.3	1,150	76	68.7	46.6
Flintshire	148.6	851	793	1,644	1.70	39.6	66.1	6.6	2,025	123	68.4	44.2
Gwynedd	116.8	604	601	1,205	1.66	48.5	73.3	7.5	1,479	84	66.4	41.1
Isle of Anglesey	66.7	327	328	655	1.75	47.3	64.2	8.4	794	41	66.4	33.1
Merthyr Tydfil	56.0	322	280	602	1.74	63.0	42.5	9.1	739	74	65.4	61.7
Monmouthshire	85.0	397	391	788	1.68	35.2	69.7	6.5	964	45	63.7	29.3
Neath Port Talbot	134.4	696	642	1,338	1.70	51.3	56.3	8.0	1,662	151	64.1	59.0
Newport	137.0	804	784	1,588	1.81	47.7	58.0	7.9	1,978	137	71.7	50.1
Pembrokeshire	113.0	649	521	1,170	1.93	47.6	66.6	8.0	1,392	100	69.3	45.0
Powys	126.4	607	575	1,182	1.82	39.9	76.5	6.8	1,382	67	63.9	30.3
Rhondda, Cynon, Taff	231.9	1,288	1,213	2,501	1.64	56.4	58.9	8.4	3,029	286	64.3	62.7
Swansea	223.2	1,189	1,208	2,397	1.67	48.9	58.9	7.6	3,024	213	67.8	51.0
Torfaen	90.9	512	456	968	1.77	51.5	61.3	7.3	1,202	94	67.5	53.7
The Vale of Glamorgan	119.3	620	578	1,198	1.68	43.3	58.6	6.3	1,474	74	63.7	30.5
Wrexham	128.5	721	700	1,421	1.67	45.9	63.8	6.3	1,824	125	70.7	51.0
Normal residence outside England and Wales	-	146	128	274	-	18.2	44.0	17.6	-	-	-	-
SCOTLAND[4]	**5,064.2**	**26,786**	**25,741**	**52,527**	**1.49**	**43.3**	**63.2**	**-**	**62,213**	**3,735**	**57.6**	**40.2**
Aberdeen City	211.9	1,063	1,034	2,097	1.25	42.5	67.9	-	2,797	183	56.5	56.0
Aberdeenshire	226.9	1,152	1,095	2,247	1.62	30.7	80.3	-	2,614	107	57.4	23.9
Angus	108.4	573	530	1,103	1.76	41.7	69.3	-	1,357	76	65.0	38.0
Argyll & Bute	91.3	399	381	780	1.54	36.5	73.7	-	875	39	53.5	23.5
Clackmannanshire	48.1	260	269	529	1.74	48.2	63.5	-	608	52	61.5	54.5
Dumfries & Galloway	147.8	661	622	1,283	1.59	42.6	64.2	-	1,532	116	56.1	43.9
Dundee City	145.5	756	712	1,468	1.39	58.2	65.1	-	1,962	167	60.8	64.8
East Ayrshire	120.3	588	610	1,198	1.55	47.5	63.3	-	1,396	97	57.1	42.2
East Dunbartonshire	108.3	535	448	983	1.46	26.7	66.4	-	961	47	43.5	20.9
East Lothian	90.2	482	458	940	1.70	35.5	69.8	-	1,088	60	60.8	37.2

Note: See Appendix A for fuller descriptions and definitions of table headings.

1 Number of live births outside marriage, which are registered by both parents who gave the same address of usual residence, expressed as a percentage of all live births outside marriage.
2 Number of live births under 2,500 grams as a percentage of all live births for which the birthweight is known.
4 Includes births to women normally resident outside Scotland. Includes deaths of persons normally resident outside Scotland.
5 The data shown in this table for Scotland may differ from that published elsewhere as these exclude miscarriages and other outcomes of pregnancy and are calculated using a different denominator.

United Kingdom by countries and, within England, Government Office Regions; counties/unitary authorities, districts and London boroughs/Northern Ireland council districts

Deaths				Deaths under 1 year				Stillbirth rate	Perinatal mortality rate	Infant mortality rate	Area
Male	Female	Total	SMR	Stillbirths	Under 1 week	Under 4 weeks	Under 1 year				
2,634	2,937	5,571	88	14	14	19	26	2.9	5.9	5.5	**Somerset**
504	577	1,081	90	1	3	3	3	*1.0*	*4.0*	*3.0*	Mendip
584	618	1,202	92	4	4	4	8	*4.1*	*8.1*	*8.1*	Sedgemoor
784	890	1,674	87	6	6	11	13	*4.0*	*8.1*	*8.8*	South Somerset
536	577	1,113	87	2	1	1	1	*2.0*	*2.9*	*1.0*	Taunton Deane
226	275	501	84	1	-	-	1	*3.9*	*3.9*	*3.9*	West Somerset
1,987	2,221	4,208	90	24	16	18	26	5.2	8.7	5.7	**Wiltshire**
304	344	648	82	6	4	5	5	*6.7*	*11.2*	*5.6*	Kennet
559	584	1,143	94	7	6	7	9	*5.1*	*9.5*	*6.6*	North Wiltshire
554	641	1,195	89	2	-	-	-	*1.8*	*1.8*	-	Salisbury
570	652	1,222	91	9	6	6	12	*7.2*	*12.0*	*9.7*	West Wiltshire
15,589	**17,404**	**32,993**	**103**	**155**	**76**	**106**	**164**	**5.0**	**7.5**	**5.4**	**WALES**
428	471	899	120	4	2	3	5	*5.5*	*8.3*	*6.9*	Blaenau Gwent
639	758	1,397	105	7	1	1	2	*4.8*	*5.5*	*1.4*	Bridgend
877	926	1,803	113	11	7	8	12	*5.5*	*9.1*	*6.1*	Caerphilly
1,397	1,508	2,905	101	14	10	15	23	*3.9*	6.7	6.4	Cardiff
1,109	1,205	2,314	108	3	-	2	6	*1.8*	*1.8*	*3.6*	Carmarthenshire
335	411	746	83	2	2	2	2	*3.5*	*7.0*	*3.5*	Ceredigion
745	890	1,635	99	6	3	3	3	*5.6*	*8.4*	*2.8*	Conwy
561	662	1,223	99	8	5	6	7	*8.5*	*13.9*	*7.5*	Denbighshire
690	757	1,447	100	6	6	8	15	*3.6*	*7.3*	*9.1*	Flintshire
607	729	1,336	95	4	2	2	2	*3.3*	*5.0*	*1.7*	Gwynedd
369	431	800	100	5	-	-	1	*7.6*	*7.6*	*1.5*	Isle of Anglesey
304	344	648	117	5	-	-	1	*8.2*	*8.2*	*1.7*	Merthyr Tydfil
468	464	932	96	2	-	1	2	2.5	*2.5*	*2.5*	Monmouthshire
806	918	1,724	111	4	7	9	11	*3.0*	*8.2*	*8.2*	Neath Port Talbot
720	738	1,458	105	8	6	6	13	*5.0*	*8.8*	*8.2*	Newport
696	675	1,371	102	6	5	6	8	*5.1*	*9.4*	*6.8*	Pembrokeshire
743	772	1,515	94	7	-	3	6	*5.9*	*5.9*	*5.1*	Powys
1,189	1,357	2,546	108	13	5	10	12	*5.2*	*7.2*	*4.8*	Rhondda, Cynon, Taff
1,189	1,392	2,581	100	17	6	10	16	*7.0*	9.5	*6.7*	Swansea
451	511	962	104	6	2	2	4	*6.2*	*8.2*	*4.1*	Torfaen
599	687	1,286	99	7	2	3	5	*5.8*	*7.5*	*4.2*	The Vale of Glamorgan
667	798	1,465	110	10	5	6	8	*7.0*	*10.5*	*5.6*	Wrexham
799	487	1,286	-	13	3	14	36	..	..	..	Normal residence outside England and Wales
27,324	**30,058**	**57,382**	**115**	**301**	**148**	**199**	**290**	**5.7**	**8.5**	**5.5**	**SCOTLAND[4]**
1,075	1,115	2,190	108	15	3	5	7	*7.1*	*8.5*	*3.3*	Aberdeen City
1,019	1,045	2,064	98	16	5	7	16	*7.1*	9.3	*7.1*	Aberdeenshire
620	714	1,334	107	3	1	2	3	*2.7*	*3.6*	*2.7*	Angus
546	588	1,134	106	3	2	3	3	*3.8*	*6.4*	*3.8*	Argyll & Bute
247	273	520	119	2	1	3	4	*3.8*	*5.6*	*7.6*	Clackmannanshire
860	917	1,777	103	8	5	6	9	*6.2*	*10.1*	*7.0*	Dumfries & Galloway
893	986	1,879	120	7	5	8	17	*4.7*	*8.1*	*11.6*	Dundee City
713	794	1,507	127	6	5	6	10	*5.0*	*9.1*	*8.3*	East Ayrshire
441	500	941	93	8	3	3	4	*8.1*	*11.1*	*4.1*	East Dunbartonshire
443	540	983	100	2	2	2	2	*2.1*	*4.2*	*2.1*	East Lothian

Table 4.2 - *continued*

Area	Estimated resident population at 30 June 2001 (thousands)	Live births							Conceptions in 2000[5]			
		Male	Female	Total	TFR[7]	% outside marriage		Proportion under 2,500 grams[2]	All ages	Under 18	All ages conception rate per 1,000 women aged 15-44	Under 18 conception rate per 1,000 women aged 15-17
						All	% jointly registered, same address[1]					
East Renfrewshire	89.4	483	471	954	1.72	23.9	68.0	-	1,033	32	57.5	18.5
Edinburgh, City of	449.0	2,280	2,209	4,489	1.19	39.8	69.9	-	6,200	322	57.4	47.2
Eilean Siar	26.5	117	109	226	1.60	26.1	47.5	-	56	5	12.0	*9.9*
Falkirk	145.3	717	731	1,448	1.46	43.2	67.5	-	1,781	110	58.5	42.3
Fife	349.8	1,860	1,782	3,642	1.59	44.9	65.3	-	4,349	278	60.5	42.7
Glasgow City	578.7	3,376	3,269	6,645	1.38	53.1	52.7	-	7,365	522	53.2	51.0
Highland	208.9	1,110	1,021	2,131	1.76	42.2	71.7	-	2,552	142	63.7	36.7
Inverclyde	84.2	416	434	850	1.54	48.2	49.8	-	1,028	70	58.5	42.0
Midlothian	81.0	464	458	922	1.77	41.8	61.8	-	1,156	78	68.6	51.1
Moray	87.0	440	429	869	1.67	34.8	69.9	-	1,062	55	63.8	32.9
North Ayrshire	135.8	707	714	1,421	1.63	52.4	56.9	-	1,632	104	57.6	38.9
North Lanarkshire	321.2	1,905	1,771	3,676	1.59	47.7	57.1	-	3,978	247	56.0	38.3
Orkney Islands	19.2	81	93	174	1.56	32.8	86.0	-	151	4	42.6	*10.3*
Perth & Kinross	135.0	691	643	1,334	1.70	35.5	74.3	-	1,690	89	66.1	35.0
Renfrewshire	172.9	911	885	1,796	1.50	45.3	57.1	-	2,191	137	58.8	43.0
Scottish Borders	107.0	529	537	1,066	1.71	36.1	76.9	-	1,174	48	58.8	26.0
Shetland Islands	22.0	132	115	247	1.85	38.1	80.9	-	230	14	53.4	*32.0*
South Ayrshire	112.2	573	483	1,056	1.59	45.1	67.0	-	1,225	73	57.0	36.3
South Lanarkshire	302.3	1,627	1,534	3,161	1.54	41.4	59.0	-	3,416	175	52.9	29.3
Stirling	86.2	437	403	840	1.40	34.0	63.3	-	1,068	53	57.2	32.3
West Dunbartonshire	93.3	479	493	972	1.51	50.4	53.1	-	1,173	82	57.8	43.6
West Lothian	159.0	982	998	1,980	1.72	43.8	67.4	-	2,370	149	66.6	49.5
Not stated	-	-	-	-	-	-	-	-	143	2	-	-
Normal residence outside Scotland	-	115	114	229	-	28.4	58.5	-	-	-	-	-
NORTHERN IRELAND[6]	**1,689.3**	**11,288**	**10,674**	**21,962**	**1.80**	**32.5**	**33.4**	**-**	**-**	**-**	**-**	**-**
Antrim	48.8	389	338	727	2.00	31.2	43.6	-	-	-	-	-
Ards	73.4	438	387	825	1.73	29.1	46.3	-	-	-	-	-
Armagh	54.5	377	380	757	2.00	21.8	35.2	-	-	-	-	-
Ballymena	58.8	361	360	721	1.82	30.7	36.7	-	-	-	-	-
Ballymoney	27.0	172	194	366	1.86	27.0	33.3	-	-	-	-	-
Banbridge	41.5	300	260	560	1.90	22.1	43.5	-	-	-	-	-
Belfast	277.2	1,677	1,621	3,298	1.53	51.1	26.8	-	-	-	-	-
Carrickfergus	37.7	245	214	459	1.71	32.7	48.0	-	-	-	-	-
Castlereagh	66.5	405	381	786	1.72	24.3	50.8	-	-	-	-	-
Coleraine	56.4	340	294	634	1.58	36.6	38.8	-	-	-	-	-
Cookstown	32.7	265	196	461	1.95	24.3	19.6	-	-	-	-	-
Craigavon	80.9	605	510	1,115	1.94	29.5	36.2	-	-	-	-	-
Derry	105.3	774	763	1,537	1.91	41.2	18.0	-	-	-	-	-
Down	64.1	419	432	851	2.00	32.4	35.1	-	-	-	-	-
Dungannon	47.8	355	358	713	2.05	22.0	24.2	-	-	-	-	-
Fermanagh	57.7	376	345	721	1.90	21.8	33.1	-	-	-	-	-
Larne	30.8	154	177	331	1.67	32.0	36.8	-	-	-	-	-
Limavady	32.6	243	242	485	1.89	28.7	30.2	-	-	-	-	-
Lisburn	109.0	777	713	1,490	1.87	34.0	39.5	-	-	-	-	-
Magherafelt	39.9	301	264	565	1.99	18.4	46.2	-	-	-	-	-
Moyle	16.0	100	96	196	1.91	30.6	40.0	-	-	-	-	-
Newry and Mourne	87.4	740	671	1,411	2.19	24.2	28.2	-	-	-	-	-
Newtownabbey	80.1	485	468	953	1.60	31.3	44.0	-	-	-	-	-
North Down	76.6	408	421	829	1.65	29.7	53.3	-	-	-	-	-
Omagh	48.1	320	332	652	1.96	23.2	34.4	-	-	-	-	-
Strabane	38.3	262	257	519	1.94	37.4	18.0	-	-	-	-	-
Normal residence outside Northern Ireland	-	160	129	289	-	12.1	48.6	-	-	-	-	-

Note: See Appendix A for fuller descriptions and definitions of table headings.

1 Number of live births outside marriage, which are registered by both parents who gave the same address of usual residence, expressed as a percentage of all live births outside marriage.
2 Number of live births under 2,500 grams as a percentage of all live births for which the birthweight is known.
5 The data shown in this table for Scotland may differ from that published elsewhere as these exclude miscarriages and other outcomes of pregnancy and are calculated using a different denominator.
6 Excludes births to women normally resident outside Northern Ireland. Includes deaths of persons normally resident outside Northern Ireland.
7 TFRs for local government districts in Northern Ireland are based on three years data rather than the single year 2001. This is due to the small numbers involved.

United Kingdom by countries and, within England, Government Office Regions; counties/unitary authorities, districts and London boroughs/Northern Ireland council districts

Deaths				Deaths under 1 year				Stillbirth rate	Perinatal mortality rate	Infant mortality rate	Area
Male	Female	Total	SMR	Stillbirths	Under 1 week	Under 4 weeks	Under 1 year				
391	424	815	93	1	3	4	5	*1.0*	*4.2*	*5.2*	East Renfrewshire
2,246	2,471	4,717	106	26	11	18	23	5.8	8.2	5.1	Edinburgh, City of
180	176	356	105	2	2	2	2	*8.8*	*17.5*	*8.8*	Eilean Siar
742	832	1,574	114	8	6	7	11	*5.5*	*9.6*	*7.6*	Falkirk
1,816	2,102	3,918	110	19	11	12	15	*5.2*	8.2	*4.1*	Fife
3,748	3,932	7,680	139	50	19	25	37	7.5	10.3	5.6	Glasgow City
1,171	1,242	2,413	111	12	9	10	11	*5.6*	9.8	*5.2*	Highland
541	568	1,109	130	4	3	4	8	*4.7*	*8.2*	*9.4*	Inverclyde
394	445	839	113	1	2	4	5	*1.1*	*3.3*	*5.4*	Midlothian
458	478	936	107	2	3	4	8	*2.3*	*5.7*	*9.2*	Moray
777	861	1,638	121	8	8	9	15	*5.6*	*11.2*	*10.6*	North Ayrshire
1,693	1,747	3,440	130	25	9	13	21	6.8	9.2	5.7	North Lanarkshire
114	115	229	109	3	-	-	1	*16.9*	*16.9*	*5.7*	Orkney Islands
703	837	1,540	97	6	3	5	6	*4.5*	*6.7*	*4.5*	Perth & Kinross
952	1,066	2,018	124	6	4	9	10	*3.3*	*5.5*	*5.6*	Renfrewshire
592	690	1,282	100	3	3	3	3	*2.8*	*5.6*	*2.8*	Scottish Borders
106	118	224	109	1	-	-	-	*4.0*	*4.0*	-	Shetland Islands
670	795	1,465	110	11	5	6	7	*10.3*	*15.0*	*6.6*	South Ayrshire
1,491	1,809	3,300	119	20	5	7	9	6.3	7.9	*2.8*	South Lanarkshire
418	521	939	110	3	2	4	6	*3.6*	*5.9*	*7.1*	Stirling
536	629	1,165	131	10	3	3	7	*10.2*	*13.2*	*7.2*	West Dunbartonshire
728	728	1,456	126	10	5	5	5	*5.0*	*7.5*	*2.5*	West Lothian
-	-	-	-	-	-	-	-	-	-	-	Not stated
270	170	440	-	2	1	2	3	..	..	..	Normal residence outside Scotland
7,007	**7,506**	**14,513**	**103**	**112**	**75**	**98**	**134**	**5.1**	**8.4**	**6.0**	**NORTHERN IRELAND[6]**
184	167	351	103	5	1	3	5	*6.8*	*8.2*	*6.9*	Antrim
325	318	643	95	8	2	3	4	*9.6*	*12.0*	*4.8*	Ards
231	230	461	111	1	1	4	5	*1.3*	*2.6*	*6.6*	Armagh
257	268	525	98	3	-	-	1	*4.1*	*4.1*	*1.4*	Ballymena
100	108	208	93	2	-	-	1	*5.4*	*5.4*	*2.7*	Ballymoney
169	165	334	100	2	4	5	5	*3.6*	*10.7*	*8.9*	Banbridge
1,355	1,532	2,887	110	22	12	15	23	6.6	10.1	6.9	Belfast
138	197	335	109	1	2	3	4	*2.2*	*6.5*	*8.7*	Carrickfergus
297	342	639	98	3	1	1	2	*3.8*	*5.1*	*2.5*	Castlereagh
238	228	466	92	6	3	4	4	*9.4*	*14.0*	*6.3*	Coleraine
130	111	241	99	-	2	3	4	-	*4.3*	*8.7*	Cookstown
319	318	637	100	5	6	8	10	*4.5*	*9.7*	*8.9*	Craigavon
330	384	714	110	7	4	5	8	*4.5*	*6.7*	*4.9*	Derry
252	282	534	101	5	2	3	7	*5.8*	*8.2*	*8.2*	Down
197	197	394	106	2	3	3	3	*2.8*	*7.0*	*4.2*	Dungannon
297	283	580	113	2	2	2	4	*2.8*	*5.5*	*5.5*	Fermanagh
147	125	272	96	4	1	1	2	*11.9*	*14.9*	*6.0*	Larne
117	98	215	103	2	1	2	2	*4.1*	6.2	*4.1*	Limavady
372	444	816	100	9	6	8	9	*6.0*	*10.0*	*6.0*	Lisburn
135	132	267	92	3	2	2	2	*5.3*	*8.8*	*3.5*	Magherafelt
73	90	163	111	2	2	2	2	*10.1*	*20.2*	*10.2*	Moyle
327	359	686	109	9	5	7	10	*6.3*	*9.1*	*6.6*	Newry and Mourne
309	353	662	97	3	6	7	8	*3.1*	*9.4*	*8.4*	Newtownabbey
338	427	765	92	1	1	1	2	*1.2*	2.4	*2.4*	North Down
202	196	398	110	1	1	1	1	*1.5*	*3.1*	*1.5*	Omagh
168	152	320	116	4	5	5	6	*7.6*	*17.2*	*11.6*	Strabane
42	36	78	-	-	3	3	3	..	..	..	Normal residence outside Northern Ireland

Table 4.3 Live births, stillbirths, total births, deaths, infant and perinatal mortality during 2001 and conceptions during 2000

Area	Estimated resident population at 30 June 2001 (thousands)	Live births							Conceptions in 2000			
		Male	Female	Total	TFR	% outside marriage: All	% outside marriage: % jointly registered, same address[1]	Proportion under 2,500 grams[2]	All ages	Under 18	All ages conception rate per 1,000 women aged 15-44	Under 18 conception rate per 1,000 women aged 15-17
UNITED KINGDOM	**58,838.0**	**342,709**	**326,414**	**669,123**	**1.63**	**40.1**	**62.4**	-	-	-	-	-
ENGLAND AND WALES	**52,084.5**	**304,635**	**289,999**	**594,634**	**1.64**	**40.0**	**63.2**	**7.6**	**766,955**	**41,348**	**71.0**	**44.1**
ENGLAND	**49,181.3**	**288,743**	**275,001**	**563,744**	**1.64**	**39.6**	**63.3**	**7.6**	**729,063**	**38,699**	**71.3**	**43.9**
Northern and Yorkshire	**6,219.7**	**34,702**	**32,733**	**67,435**	**1.63**	**44.7**	**61.8**	**7.8**	**83,108**	**5,552**	**65.2**	**47.5**
Bradford	467.9	3,743	3,462	7,205	2.18	32.8	61.3	9.6	8,354	499	84.6	49.6
Calderdale and Kirklees	581.4	3,712	3,586	7,298	1.86	38.8	65.9	9.0	8,625	475	72.0	43.6
County Durham and Darlington	591.6	3,057	2,789	5,846	1.54	51.1	61.8	7.2	7,088	529	59.1	49.6
East Riding and Hull	558.2	2,928	2,721	5,649	1.61	51.5	66.5	7.8	7,121	524	64.7	51.3
Gateshead and South Tyneside	344.0	1,825	1,649	3,474	1.60	53.0	54.3	7.3	4,288	373	61.1	57.1
Leeds	715.5	4,034	3,797	7,831	1.43	44.0	63.7	8.1	9,941	631	62.8	49.1
Newcastle & North Tyneside	451.6	2,491	2,333	4,824	1.48	47.7	54.1	7.2	6,300	458	64.2	57.3
North Cumbria	313.4	1,485	1,476	2,961	1.58	42.4	65.9	6.6	3,718	253	61.9	44.8
Northumberland	307.4	1,433	1,448	2,881	1.65	42.3	66.9	6.1	3,545	238	60.9	40.9
North Yorkshire	751.4	3,797	3,464	7,261	1.54	34.4	69.0	5.9	8,866	404	61.8	31.1
Sunderland	280.8	1,453	1,400	2,853	1.50	56.7	55.0	7.7	3,798	290	64.0	51.0
Tees	541.3	3,121	2,950	6,071	1.73	53.5	54.4	8.5	7,448	590	66.0	51.4
Wakefield	315.4	1,623	1,658	3,281	1.59	48.7	66.5	8.2	4,016	288	62.0	48.9
Trent	**5,089.6**	**27,861**	**26,148**	**54,009**	**1.61**	**44.9**	**66.3**	**7.7**	**67,302**	**4,220**	**65.4**	**46.3**
Barnsley	218.1	1,127	1,099	2,226	1.64	53.1	65.1	7.3	2,762	207	62.7	52.5
Doncaster	286.9	1,640	1,585	3,225	1.85	52.3	65.0	8.3	4,054	386	70.9	70.0
Leicestershire	924.7	5,432	5,071	10,503	1.63	36.8	67.9	8.0	13,174	663	68.2	38.8
Lincolnshire	647.6	3,128	2,892	6,020	1.61	43.2	72.2	6.9	7,428	446	62.4	40.4
North Derbyshire	367.7	1,759	1,661	3,420	1.58	44.6	72.4	6.9	4,253	245	60.5	38.7
North Nottinghamshire	394.0	1,980	1,959	3,939	1.63	49.8	69.2	7.9	4,766	316	61.5	44.2
Nottingham	621.7	3,457	3,268	6,725	1.48	47.0	59.0	8.0	8,485	563	63.4	51.6
Rotherham	248.4	1,429	1,301	2,730	1.74	48.8	60.5	7.7	3,375	238	67.7	51.2
Sheffield	513.1	2,939	2,740	5,679	1.53	44.3	62.9	8.2	7,239	424	66.2	49.4
Southern Derbyshire	556.4	3,308	2,994	6,302	1.69	42.6	69.8	7.6	7,582	402	66.8	40.5
South Humber	310.9	1,662	1,578	3,240	1.75	53.4	65.1	7.8	4,184	330	68.8	55.1
Eastern	**5,394.9**	**30,729**	**29,361**	**60,090**	**1.67**	**35.9**	**71.0**	**6.7**	**74,492**	**3,331**	**68.9**	**35.3**
Bedfordshire	566.4	3,754	3,581	7,335	1.82	31.6	65.5	8.3	9,218	407	76.6	37.9
Cambridgeshire	710.0	4,063	3,853	7,916	1.55	34.5	70.5	7.1	9,752	428	64.8	34.2
Hertfordshire	1,034.9	6,466	6,228	12,694	1.70	31.4	72.2	6.3	15,728	593	72.7	32.5
Norfolk	797.9	3,756	3,649	7,405	1.56	42.4	72.9	6.8	9,249	496	63.0	37.1
North Essex	913.0	5,024	4,661	9,685	1.63	35.7	72.3	6.2	12,070	528	67.9	34.1
South Essex	703.2	4,110	3,989	8,099	1.72	42.4	69.3	6.2	10,225	516	72.1	42.2
Suffolk	669.4	3,556	3,400	6,956	1.72	35.7	72.8	6.5	8,250	363	64.7	30.7

Notes: See Appendix A for fuller descriptions and definitions of table headings.
Figures are displayed for health authority boundaries as at 1 April 2001.

1 Number of live births outside marriage, which are registered by both parents who gave the same address of usual residence, expressed as a percentage of all live births outside marriage.
2 Number of live births under 2,500 grams as a percentage of all live births for which the birthweight is known.

United Kingdom by countries and, within England, Health Regional Office areas; health authorities

Deaths				Deaths under 1 year				Stillbirth rate	Perinatal mortality rate	Infant mortality rate	Area
Male	Female	Total	SMR	Stillbirths	Under 1 week	Under 4 weeks	Under 1 year				
286,757	**315,511**	**602,268**	**100**	**3,572**	**1,821**	**2,434**	**3,664**	**5.3**	**8.0**	**5.5**	**UNITED KINGDOM**
252,426	**277,947**	**530,373**	**99**	**3,159**	**1,598**	**2,137**	**3,240**	**5.3**	**8.0**	**5.4**	**ENGLAND AND WALES**
236,038	**260,056**	**496,094**	**98**	**2,991**	**1,519**	**2,017**	**3,040**	**5.3**	**8.0**	**5.4**	**ENGLAND**
31,794	**34,665**	**66,459**	**105**	**343**	**159**	**220**	**362**	**5.1**	**7.4**	**5.4**	**Northern and Yorkshire**
2,213	2,502	4,715	108	55	23	30	63	7.6	10.7	8.7	Bradford
2,743	3,202	5,945	104	33	21	24	43	4.5	7.4	5.9	Calderdale and Kirklees
3,203	3,477	6,680	110	23	16	21	39	3.9	6.6	6.7	County Durham and Darlington
2,870	3,061	5,931	99	20	12	20	30	3.5	5.6	5.3	East Riding and Hull
2,022	2,044	4,066	112	25	9	12	18	7.1	9.7	*5.2*	Gateshead and South Tyneside
3,321	3,628	6,949	98	48	14	22	39	6.1	7.9	5.0	Leeds
2,411	2,763	5,174	109	19	11	15	22	*3.9*	6.2	4.6	Newcastle & North Tyneside
1,692	1,842	3,534	102	14	4	7	10	*4.7*	*6.1*	*3.4*	North Cumbria
1,673	1,796	3,469	103	19	3	6	10	*6.6*	7.6	*3.5*	Northumberland
3,742	4,162	7,904	90	26	10	14	17	3.6	4.9	*2.3*	North Yorkshire
1,539	1,555	3,094	119	16	7	9	12	*5.6*	8.0	*4.2*	Sunderland
2,734	2,914	5,648	111	34	21	29	38	5.6	9.0	6.3	Tees
1,631	1,719	3,350	110	11	8	11	21	*3.3*	*5.8*	6.4	Wakefield
25,763	**27,883**	**53,646**	**102**	**293**	**144**	**190**	**285**	**5.4**	**8.0**	**5.3**	**Trent**
1,207	1,281	2,488	113	8	5	5	10	*3.6*	*5.8*	*4.5*	Barnsley
1,443	1,559	3,002	105	18	3	6	14	*5.6*	6.5	*4.3*	Doncaster
4,057	4,544	8,601	95	65	33	43	55	6.2	9.3	5.2	Leicestershire
3,588	3,708	7,296	96	28	10	15	26	4.6	6.3	4.3	Lincolnshire
2,002	2,262	4,264	103	22	6	9	14	6.4	8.1	*4.1*	North Derbyshire
2,138	2,282	4,420	108	13	9	11	20	*3.3*	5.6	5.1	North Nottinghamshire
2,997	3,359	6,356	103	29	22	30	45	4.3	7.6	6.7	Nottingham
1,245	1,322	2,567	106	11	8	9	15	*4.0*	*6.9*	*5.5*	Rotherham
2,666	2,935	5,601	102	39	19	24	34	6.8	10.1	6.0	Sheffield
2,772	2,964	5,736	100	42	19	26	32	6.6	9.6	5.1	Southern Derbyshire
1,648	1,667	3,315	102	18	10	12	20	*5.5*	8.6	6.2	South Humber
25,264	**28,023**	**53,287**	**93**	**291**	**136**	**176**	**271**	**4.8**	**7.1**	**4.5**	**Eastern**
2,347	2,519	4,866	99	51	20	24	41	6.9	9.6	5.6	Bedfordshire
3,093	3,237	6,330	92	33	19	22	39	4.2	6.5	4.9	Cambridgeshire
4,386	5,067	9,453	93	66	32	42	60	5.2	7.7	4.7	Hertfordshire
4,420	4,700	9,120	89	35	20	26	33	4.7	7.4	4.5	Norfolk
4,354	4,976	9,330	93	41	17	27	41	4.2	6.0	4.2	North Essex
3,215	3,779	6,994	95	40	11	18	31	4.9	6.3	3.8	South Essex
3,449	3,745	7,194	90	25	17	17	26	3.6	6.0	3.7	Suffolk

Table 4.3 - *continued*

Area	Estimated resident population at 30 June 2001 (thousands)	Live births							Conceptions in 2000			
		Male	Female	Total	TFR	% outside marriage		Proportion under 2,500 grams[2]	All ages	Under 18	All ages conception rate per 1,000 women aged 15-44	Under 18 conception rate per 1,000 women aged 15-17
						All	% jointly registered, same address[1]					
London	**7,188.0**	**53,157**	**51,005**	**104,162**	**1.62**	**34.6**	**53.1**	**8.1**	**153,533**	**6,041**	**87.8**	**50.8**
Barking & Havering	389.1	2,486	2,303	4,789	1.75	45.0	62.5	6.8	6,407	372	78.7	50.5
Barnet, Enfield and Haringey	806.4	5,998	5,642	11,640	1.66	32.7	52.0	7.8	16,978	638	87.6	45.4
Bexley, Bromley and Greenwich	730.1	4,663	4,587	9,250	1.68	40.4	60.8	7.6	12,333	521	78.5	42.0
Brent & Harrow	471.8	3,292	3,206	6,498	1.60	26.7	49.5	8.8	10,143	343	90.5	39.1
Camden & Islington	374.5	2,687	2,605	5,292	1.36	37.2	50.5	7.4	8,583	306	83.3	56.6
Croydon	331.5	2,216	2,185	4,401	1.67	41.5	50.5	8.8	6,442	347	84.0	56.0
Ealing, Hammersmith & Hounslow	679.7	5,150	4,741	9,891	1.55	28.5	51.4	7.8	14,551	489	84.4	46.0
East London and The City	651.5	6,394	6,205	12,599	2.02	31.6	45.0	9.9	18,309	808	109.7	60.6
Hillingdon	243.1	1,662	1,582	3,244	1.67	34.5	60.3	7.7	4,503	181	81.3	41.7
Kensington & Chelsea and Westminster	340.8	2,346	2,362	4,708	1.38	23.4	52.5	7.6	7,601	199	88.2	62.3
Kingston and Richmond	320.4	2,093	2,088	4,181	1.50	23.8	70.8	6.3	5,435	127	74.7	28.0
Lambeth, Southwark and Lewisham	761.7	6,159	5,954	12,113	1.63	49.6	45.9	9.4	20,363	947	100.1	80.6
Merton, Sutton and Wandsworth	629.4	4,615	4,321	8,936	1.47	31.2	60.5	6.9	12,446	436	77.2	50.7
Redbridge and Waltham Forest	458.0	3,396	3,224	6,620	1.75	30.9	54.7	8.5	9,439	327	89.0	39.7
South East	**8,637.3**	**49,221**	**46,673**	**95,894**	**1.63**	**35.3**	**70.5**	**6.9**	**120,379**	**5,610**	**68.6**	**36.8**
Berkshire	800.3	5,099	4,921	10,020	1.64	30.2	67.9	7.5	12,892	542	72.8	37.3
Buckinghamshire	686.8	4,269	4,024	8,293	1.73	31.1	71.4	7.3	10,315	453	71.3	35.4
East Kent	592.2	3,164	2,988	6,152	1.69	46.0	68.2	6.9	7,709	546	68.4	50.8
East Surrey	421.3	2,324	2,268	4,592	1.63	25.8	76.5	6.1	5,720	180	69.7	25.5
East Sussex, Brighton and Hove	741.2	3,781	3,614	7,395	1.52	42.1	72.2	6.5	9,903	512	68.9	42.1
Isle of Wight, Portsmouth and South East Hampshire	674.8	3,512	3,362	6,874	1.60	41.5	66.8	7.6	8,639	489	64.9	39.8
Northamptonshire	630.4	3,764	3,620	7,384	1.75	41.5	68.9	7.2	9,089	525	70.1	45.5
North and Mid Hampshire	554.7	3,241	2,965	6,206	1.62	30.6	72.9	6.1	7,489	266	65.8	27.0
Oxfordshire	605.9	3,494	3,525	7,019	1.54	29.8	71.3	6.8	8,784	320	66.5	30.0
Southampton and South West Hampshire	548.7	2,876	2,640	5,516	1.47	38.6	68.4	7.1	6,818	367	61.1	39.9
West Kent	988.7	5,943	5,577	11,520	1.76	39.3	70.2	7.0	14,385	724	72.0	38.7
West Surrey	638.2	3,770	3,400	7,170	1.54	26.6	75.3	6.3	9,096	275	68.6	26.2
West Sussex	754.3	3,984	3,769	7,753	1.68	33.9	72.6	6.9	9,540	411	67.4	33.1
South West	**4,934.2**	**24,938**	**23,809**	**48,747**	**1.59**	**39.3**	**71.1**	**6.7**	**60,231**	**3,092**	**64.0**	**36.3**
Avon	984.7	5,567	5,342	10,909	1.53	39.5	68.6	6.7	13,517	657	65.2	39.3
Cornwall and Isles of Scilly	502.1	2,298	2,179	4,477	1.65	44.1	71.9	6.5	5,505	309	62.7	35.4
Dorset	693.5	3,145	3,025	6,170	1.52	37.3	70.9	6.7	7,658	362	61.3	31.6
Gloucestershire	565.0	2,964	2,818	5,782	1.61	38.4	71.5	6.8	7,211	353	65.0	35.3
North and East Devon	487.9	2,246	2,153	4,399	1.55	40.4	72.2	6.1	5,335	254	60.7	30.9
Somerset	498.7	2,415	2,317	4,732	1.68	37.9	73.0	6.5	5,753	333	63.8	36.6
South and West Devon	588.6	2,808	2,636	5,444	1.58	44.2	69.2	6.4	6,857	432	62.4	42.0
Wiltshire	613.7	3,495	3,339	6,834	1.71	35.1	74.2	7.2	8,395	392	68.4	37.0

Notes: See Appendix A for fuller descriptions and definitions of table headings.
Figures are displayed for health authority boundaries as at 1 April 2001.

1 Number of live births outside marriage, which are registered by both parents who gave the same address of usual residence, expressed as a percentage of all live births outside marriage.
2 Number of live births under 2,500 grams as a percentage of all live births for which the birthweight is known.

United Kingdom by countries and, within England, Health Regional Office areas; health authorities

Deaths: Male	Deaths: Female	Deaths: Total	Deaths: SMR	Deaths under 1 year: Stillbirths	Deaths under 1 year: Under 1 week	Deaths under 1 year: Under 4 weeks	Deaths under 1 year: Under 1 year	Stillbirth rate	Perinatal mortality rate	Infant mortality rate	Area
28,308	**29,858**	**58,166**	**98**	**620**	**314**	**428**	**637**	**5.9**	**8.9**	**6.1**	**London**
1,958	2,082	4,040	101	23	7	10	15	4.8	6.2	*3.1*	Barking & Havering
3,070	3,604	6,674	94	86	32	45	62	7.3	10.1	5.3	Barnet, Enfield and Haringey
3,268	3,682	6,950	95	51	24	33	46	5.5	8.1	5.0	Bexley, Bromley and Greenwich
1,696	1,712	3,408	85	37	25	31	44	5.7	9.5	6.8	Brent & Harrow
1,492	1,340	2,832	108	34	17	20	32	6.4	9.6	6.0	Camden & Islington
1,225	1,420	2,645	92	26	9	12	19	5.9	7.9	*4.3*	Croydon
2,508	2,504	5,012	97	53	32	45	68	5.3	8.5	6.9	Ealing, Hammersmith & Hounslow
2,341	1,998	4,339	109	76	44	58	84	6.0	9.5	6.7	East London and The City
989	1,089	2,078	94	16	10	13	21	*4.9*	8.0	6.5	Hillingdon
1,185	1,152	2,337	82	27	13	18	31	5.7	8.4	6.6	Kensington & Chelsea and Westminster
1,272	1,507	2,779	90	25	9	14	21	5.9	8.1	5.0	Kingston and Richmond
2,896	2,800	5,696	110	81	49	65	100	6.6	10.7	8.3	Lambeth, Southwark and Lewisham
2,474	2,878	5,352	102	39	14	23	37	4.3	5.9	4.1	Merton, Sutton and Wandsworth
1,934	2,090	4,024	100	46	29	41	57	6.9	11.3	8.6	Redbridge and Waltham Forest
39,568	**45,429**	**84,997**	**92**	**446**	**214**	**277**	**402**	**4.6**	**6.9**	**4.2**	**South East**
3,106	3,236	6,342	94	48	35	41	56	4.8	8.2	5.6	Berkshire
2,677	2,942	5,619	92	44	38	44	55	5.3	9.8	6.6	Buckinghamshire
3,300	3,851	7,151	95	23	15	17	27	3.7	6.2	4.4	East Kent
2,011	2,328	4,339	90	15	8	9	13	*3.3*	5.0	*2.8*	East Surrey
4,397	5,195	9,592	92	31	15	22	32	4.2	6.2	4.3	East Sussex, Brighton and Hove
3,448	3,998	7,446	95	49	12	24	35	7.1	8.8	5.1	Isle of Wight, Portsmouth and South East Hampshire
2,686	2,863	5,549	94	37	13	16	26	5.0	6.7	3.5	Northamptonshire
2,112	2,412	4,524	88	22	10	16	29	3.5	5.1	4.7	North and Mid Hampshire
2,458	2,638	5,096	87	34	15	17	28	4.8	6.9	4.0	Oxfordshire
2,433	2,717	5,150	83	26	11	16	24	4.7	6.7	4.4	Southampton and South West Hampshire
4,230	4,830	9,060	96	62	21	25	34	5.4	7.2	3.0	West Kent
2,598	3,167	5,765	86	19	10	14	22	*2.6*	4.0	3.1	West Surrey
4,112	5,252	9,364	91	36	11	16	21	4.6	6.0	2.7	West Sussex
25,620	**28,667**	**54,287**	**90**	**211**	**142**	**180**	**263**	**4.3**	**7.2**	**5.4**	**South West**
4,532	5,047	9,579	92	43	33	43	68	3.9	6.9	6.2	Avon
2,822	3,109	5,931	90	14	12	14	23	*3.1*	5.8	5.1	Cornwall and Isles of Scilly
4,049	4,598	8,647	85	25	19	22	31	4.0	7.1	5.0	Dorset
2,766	3,063	5,829	90	31	14	21	27	5.3	7.7	4.7	Gloucestershire
2,830	3,122	5,952	89	19	12	17	26	*4.3*	7.0	5.9	North and East Devon
2,634	2,937	5,571	88	14	14	19	26	*2.9*	5.9	5.5	Somerset
3,202	3,781	6,983	92	24	13	15	21	4.4	6.8	3.9	South and West Devon
2,785	3,010	5,795	93	41	25	29	41	6.0	9.6	6.0	Wiltshire

Table 4.3 - *continued*

Area	Estimated resident population at 30 June 2001 (thousands)	Live births							Conceptions in 2000			
		Male	Female	Total	TFR	% outside marriage		Proportion under 2,500 grams[2]	All ages	Under 18	All ages conception rate per 1,000 women aged 15-44	Under 18 conception rate per 1,000 women aged 15-17
						All	% jointly registered, same address[1]					
West Midlands	**5,267.1**	**30,896**	**29,922**	**60,818**	**1.74**	**40.3**	**62.0**	**8.6**	**77,919**	**4,862**	**73.2**	**49.0**
Birmingham	976.4	7,358	7,068	14,426	1.96	37.1	51.5	9.9	18,724	1,133	87.2	55.9
Coventry	300.7	1,848	1,766	3,614	1.68	45.2	60.3	8.8	5,134	374	80.3	64.4
Dudley	305.1	1,724	1,589	3,313	1.70	40.7	65.3	7.8	4,271	272	71.3	50.9
Herefordshire	174.9	783	808	1,591	1.63	38.2	70.3	6.7	1,921	117	61.4	39.0
North Staffordshire	457.0	2,392	2,331	4,723	1.56	47.4	66.8	8.3	5,803	419	62.1	49.5
Sandwell	282.8	1,854	1,840	3,694	1.92	43.5	53.6	10.6	4,845	346	82.2	63.3
Shropshire	441.9	2,283	2,221	4,504	1.66	41.5	69.6	7.2	5,587	323	66.2	39.8
Solihull	199.6	1,038	921	1,959	1.63	38.8	62.4	6.9	2,519	120	66.5	32.2
South Staffordshire	590.6	3,036	3,043	6,079	1.64	37.7	72.3	8.1	7,586	442	65.4	40.0
Walsall	253.3	1,621	1,578	3,199	1.99	43.9	61.5	9.7	3,926	306	77.2	62.7
Warwickshire	506.2	2,642	2,612	5,254	1.60	36.2	68.0	7.3	6,914	373	70.4	41.9
Wolverhampton	236.4	1,457	1,443	2,900	1.78	47.9	52.3	10.2	3,788	286	77.3	63.5
Worcestershire	542.2	2,860	2,702	5,562	1.62	37.6	69.9	7.6	6,901	351	65.4	36.2
North West	**6,450.5**	**37,239**	**35,350**	**72,589**	**1.67**	**46.5**	**58.1**	**8.0**	**92,099**	**5,991**	**69.2**	**48.1**
Bury and Rochdale	385.9	2,413	2,207	4,620	1.83	43.7	61.6	9.0	5,735	386	72.2	51.0
East Lancashire	517.6	3,308	3,200	6,508	1.96	40.5	64.4	9.6	7,977	540	76.3	51.8
Liverpool	439.2	2,545	2,370	4,915	1.47	61.3	42.3	8.4	6,750	443	66.2	49.3
Manchester	392.9	2,809	2,687	5,496	1.62	52.9	47.1	8.2	7,721	524	80.7	70.2
Morecambe Bay	308.4	1,511	1,437	2,948	1.51	45.2	66.2	7.0	3,614	244	59.3	42.7
North Cheshire	309.4	1,843	1,745	3,588	1.74	47.7	59.8	6.2	4,598	307	70.5	49.2
North-West Lancashire	451.0	2,368	2,242	4,610	1.65	47.4	62.8	8.4	5,931	443	68.2	54.1
St Helens and Knowsley	327.3	1,868	1,718	3,586	1.66	57.6	47.1	8.3	4,506	343	65.4	50.5
Salford and Trafford	426.0	2,427	2,475	4,902	1.68	45.9	58.5	7.5	6,283	357	70.9	43.2
Sefton	282.9	1,324	1,316	2,640	1.57	48.4	54.1	7.6	3,336	220	61.3	39.9
South Cheshire	674.2	3,515	3,406	6,921	1.59	35.5	68.8	6.6	8,468	405	64.1	33.6
South Lancashire	313.0	1,703	1,549	3,252	1.64	41.1	67.0	7.8	3,867	220	62.4	36.2
Stockport	284.6	1,509	1,394	2,903	1.56	40.8	65.4	6.8	3,757	173	65.7	32.7
West Pennine	463.0	3,068	2,807	5,875	1.91	45.0	61.9	9.1	7,164	495	74.8	54.5
Wigan and Bolton	562.8	3,330	3,187	6,517	1.74	46.6	63.0	8.5	8,138	582	70.5	55.6
Wirral	312.2	1,698	1,610	3,308	1.74	51.4	49.7	6.7	4,254	309	69.0	48.3
WALES	**2,903.2**	**15,746**	**14,870**	**30,616**	**1.66**	**48.3**	**60.8**	**7.5**	**37,892**	**2,649**	**66.3**	**48.4**
North Wales	663.5	3,508	3,413	6,921	1.71	46.1	67.0	6.9	8,649	545	69.1	45.4
Dyfed Powys	488.3	2,446	2,142	4,588	1.68	45.3	69.8	7.1	5,566	341	62.8	37.5
Morgannwg	486.3	2,594	2,585	5,179	1.70	49.4	58.3	7.4	6,442	489	67.2	53.7
Bro Taf	712.4	4,113	3,777	7,890	1.53	49.7	55.1	7.8	9,816	700	63.9	50.4
Gwent	552.5	3,085	2,953	6,038	1.78	50.2	57.6	8.0	7,419	574	68.5	53.6
Normal residence outside England and Wales	-	146	128	274	-	18.2	44.0	17.6	-	-	-	-

Notes: See Appendix A for fuller descriptions and definitions of table headings.
Figures are displayed for health authority boundaries as at 1 April 2001.

1 Number of live births outside marriage, which are registered by both parents who gave the same address of usual residence, expressed as a percentage of all live births outside marriage.

2 Number of live births under 2,500 grams as a percentage of all live births for which the birthweight is known.

United Kingdom by countries and, within England, Health Regional Office areas; health authorities

Deaths				Deaths under 1 year				Stillbirth rate	Perinatal mortality rate	Infant mortality rate	Area
Male	Female	Total	SMR	Stillbirths	Under 1 week	Under 4 weeks	Under 1 year				
26,036	**27,829**	**53,865**	**101**	**337**	**217**	**269**	**389**	**5.5**	**9.1**	**6.4**	**West Midlands**
4,851	4,874	9,725	106	93	65	83	127	6.4	10.9	8.8	Birmingham
1,442	1,477	2,919	99	21	14	15	27	5.8	9.6	7.5	Coventry
1,568	1,610	3,178	101	20	7	10	16	6.0	8.1	*4.8*	Dudley
907	969	1,876	88	9	4	4	4	*5.6*	*8.1*	*2.5*	Herefordshire
2,379	2,604	4,983	105	22	22	25	32	4.6	9.3	6.8	North Staffordshire
1,609	1,716	3,325	116	23	9	13	25	6.2	8.6	6.8	Sandwell
2,065	2,291	4,356	95	28	19	21	27	6.2	10.4	6.0	Shropshire
898	970	1,868	89	13	6	8	9	*6.6*	*9.6*	*4.6*	Solihull
2,773	3,083	5,856	103	34	18	25	32	5.6	8.5	5.3	South Staffordshire
1,225	1,314	2,539	102	17	11	12	18	*5.3*	8.7	*5.6*	Walsall
2,418	2,716	5,134	97	18	9	15	24	*3.4*	5.1	4.6	Warwickshire
1,285	1,245	2,530	102	16	18	20	26	*5.5*	11.7	9.0	Wolverhampton
2,616	2,960	5,576	96	23	15	18	22	4.1	6.8	4.0	Worcestershire
33,685	**37,702**	**71,387**	**110**	**450**	**193**	**277**	**431**	**6.2**	**8.8**	**5.9**	**North West**
1,863	2,182	4,045	112	36	17	21	28	7.7	11.4	6.1	Bury and Rochdale
2,453	2,987	5,440	109	46	19	28	48	7.0	9.9	7.4	East Lancashire
2,457	2,718	5,175	126	26	16	23	33	5.3	8.5	6.7	Liverpool
2,251	2,222	4,473	133	44	16	26	53	7.9	10.8	9.6	Manchester
1,781	1,886	3,667	98	16	13	20	25	*5.4*	9.8	8.5	Morecambe Bay
1,473	1,571	3,044	112	16	9	10	15	*4.4*	6.9	*4.2*	North Cheshire
2,777	3,127	5,904	108	31	13	17	22	6.7	9.5	4.8	North-West Lancashire
1,654	1,812	3,466	118	22	12	13	25	6.1	9.4	7.0	St Helens and Knowsley
2,254	2,525	4,779	109	27	14	18	25	5.5	8.3	5.1	Salford and Trafford
1,607	1,943	3,550	107	11	7	7	9	*4.1*	*6.8*	*3.4*	Sefton
3,297	3,676	6,973	97	47	9	19	27	6.7	8.0	3.9	South Cheshire
1,500	1,716	3,216	106	8	7	10	15	*2.5*	*4.6*	*4.6*	South Lancashire
1,391	1,549	2,940	98	14	12	14	21	*4.8*	8.9	7.2	Stockport
2,352	2,552	4,904	112	39	8	16	33	6.6	7.9	5.6	West Pennine
2,805	3,227	6,032	115	42	16	24	38	6.4	8.8	5.8	Wigan and Bolton
1,770	2,009	3,779	106	25	5	11	14	7.5	9.0	*4.2*	Wirral
15,589	**17,404**	**32,993**	**103**	**155**	**76**	**106**	**164**	**5.0**	**7.5**	**5.4**	**WALES**
3,639	4,267	7,906	100	39	21	25	36	5.6	8.6	5.2	North Wales
2,883	3,063	5,946	99	18	7	13	22	*3.9*	5.4	4.8	Dyfed Powys
2,634	3,068	5,702	104	28	14	20	29	5.4	8.1	5.6	Morgannwg
3,489	3,896	7,385	105	39	17	28	41	4.9	7.1	5.2	Bro Taf
2,944	3,110	6,054	107	31	17	20	36	5.1	7.9	6.0	Gwent
799	487	1,286	..	13	3	14	36	..	..	..	Normal residence outside England and Wales

Table 4.3 - *continued*

Area	Estimated resident population at 30 June 2001 (thousands)	Live births: Male	Live births: Female	Live births: Total	TFR	% outside marriage: All	% outside marriage: % jointly registered, same address[1]	Proportion under 2,500 grams[2]	Conceptions in 2000[5]: All ages	Conceptions in 2000[5]: Under 18	All ages conception rate per 1,000 women aged 15-44	Under 18 conception rate per 1,000 women aged 15-17
SCOTLAND[3]	**5,064.2**	**26,786**	**25,741**	**52,527**	**1.49**	**43.3**	**63.2**	**-**	**62,213**	**3,735**	**57.6**	**40.2**
Argyll & Clyde	420.7	2,080	2,056	4,136	1.51	45.0	58.4	-	4,948	304	57.1	38.0
Ayrshire & Arran	368.3	1,868	1,807	3,675	1.60	48.7	61.6	-	4,262	274	57.5	39.3
Borders	107.0	529	537	1,066	1.71	36.1	76.9	-	1,177	48	58.9	26.0
Dumfries & Galloway	147.8	661	622	1,283	1.59	42.6	64.2	-	1,532	116	55.9	43.8
Fife	349.7	1,860	1,782	3,642	1.59	44.9	65.3	-	4,353	278	60.6	42.7
Forth Valley	279.2	1,414	1,403	2,817	1.47	41.4	65.6	-	3,467	215	58.7	41.4
Grampian	525.9	2,655	2,558	5,213	1.42	36.1	72.8	-	6,479	345	57.8	36.7
Greater Glasgow	868.2	4,944	4,712	9,656	1.44	46.8	54.0	-	10,445	659	52.9	41.2
Highland	208.9	1,110	1,021	2,131	1.76	42.2	71.7	-	2,555	142	63.7	36.9
Lanarkshire	553.2	3,107	2,918	6,025	1.56	44.9	58.3	-	6,705	389	55.7	35.3
Lothian	779.0	4,208	4,123	8,331	1.35	40.5	68.3	-	10,840	610	60.8	46.9
Orkney	19.2	81	93	174	1.56	32.8	86.0	-	151	4	42.5	*10.3*
Shetland	22.0	132	115	247	1.85	38.1	80.9	-	230	14	53.5	*32.0*
Tayside	388.8	2,020	1,885	3,905	1.57	45.8	68.6	-	5,013	332	63.4	46.6
Western Isles	26.5	117	109	226	1.60	26.1	47.5	-	56	5	12.1	*9.9*
Normal residence outside Scotland	-	115	114	229	-	28.4	58.5	-	-	-	-	-
NORTHERN IRELAND[4]	**1,689.3**	**11,288**	**10,674**	**21,962**	**1.80**	**32.5**	**33.4**	**-**	**-**	**-**	**-**	**-**
Eastern	666.9	4,124	3,955	8,079	1.66	38.9	34.6	-	-	-	-	-
Northern	428.2	2,812	2,601	5,413	1.77	29.7	39.7	-	-	-	-	-
Southern	312.2	2,377	2,179	4,556	2.07	24.5	32.7	-	-	-	-	-
Western	282.0	1,975	1,939	3,914	1.91	32.6	23.1	-	-	-	-	-
Normal residence outside Northern Ireland	-	160	129	289	-	12.1	48.6	-	-	-	-	-

Notes: See Appendix A for fuller descriptions and definitions of table headings.
Figures are displayed for health authority boundaries as at 1 April 2001.

1 Number of live births outside marriage, which are registered by both parents who gave the same address of usual residence, expressed as a percentage of all live births outside marriage.
2 Number of live births under 2,500 grams as a percentage of all live births for which the birthweight is known.
3 Includes births to women normally resident outside Scotland. Includes deaths of persons normally resident outside Scotland.
4 Excludes births to women normally resident outside Northern Ireland. Includes deaths of persons normally resident outside Northern Ireland.
5 The data shown in this table for Scotland may differ from that published elsewhere as these exclude miscarriages and other outcomes of pregnancy and are calculated using a different denominator.

United Kingdom by countries and, within England, Health Regional Office areas; health authorities

Deaths				Deaths under 1 year				Stillbirth rate	Perinatal mortality rate	Infant mortality rate	Area
Male	Female	Total	SMR	Stillbirths	Under 1 week	Under 4 weeks	Under 1 year				
27,324	**30,058**	**57,382**	**115**	**301**	**148**	**199**	**290**	**5.7**	**8.5**	**5.5**	**SCOTLAND**[3]
2,451	2,643	5,094	122	15	12	19	26	*3.6*	6.5	6.3	Argyll & Clyde
2,160	2,450	4,610	119	25	18	21	32	6.8	11.6	8.7	Ayrshire & Arran
592	690	1,282	100	3	3	3	3	*2.8*	*5.6*	*2.8*	Borders
860	917	1,777	103	8	5	6	9	*6.2*	*10.1*	*7.0*	Dumfries & Galloway
1,816	2,102	3,918	110	19	11	12	15	*5.2*	8.2	*4.1*	Fife
1,407	1,626	3,033	114	13	9	14	21	*4.6*	7.8	7.5	Forth Valley
2,552	2,638	5,190	103	33	11	16	31	6.3	8.4	5.9	Grampian
5,095	5,549	10,644	128	73	26	34	51	7.5	10.2	5.3	Greater Glasgow
1,171	1,242	2,413	111	12	9	10	11	*5.6*	9.8	*5.2*	Highland
2,793	3,071	5,864	123	39	13	18	27	6.4	8.6	4.5	Lanarkshire
3,811	4,184	7,995	109	39	20	29	35	4.7	7.0	4.2	Lothian
114	115	229	109	3	-	-	1	*16.9*	*16.9*	*5.7*	Orkney
106	118	224	109	1	-	-	-	*4.0*	*4.0*	-	Shetland
2,216	2,537	4,753	108	16	9	15	26	*4.1*	6.4	6.7	Tayside
180	176	356	105	2	2	2	2	*8.8*	*17.5*	*8.8*	Western Isles
270	170	440	..	2	1	2	3	..	..	..	Normal residence outside Scotland
7,007	**7,506**	**14,513**	**103**	**112**	**75**	**98**	**134**	**5.1**	**8.4**	**6.0**	**NORTHERN IRELAND**[4]
2,939	3,345	6,284	103	48	24	31	47	5.9	8.8	5.8	Eastern
1,711	1,779	3,490	98	29	19	25	33	5.3	8.8	6.1	Northern
1,243	1,269	2,512	105	19	19	27	33	*4.2*	8.1	7.0	Southern
1,114	1,113	2,227	111	16	13	15	21	*4.1*	7.2	5.2	Western
42	36	78	..	-	3	3	3	..	..	..	Normal residence outside Northern Ireland

Map 5a Migration during the year ending June 2001: counties and unitary authorities of England and Wales; Scotland and Northern Ireland

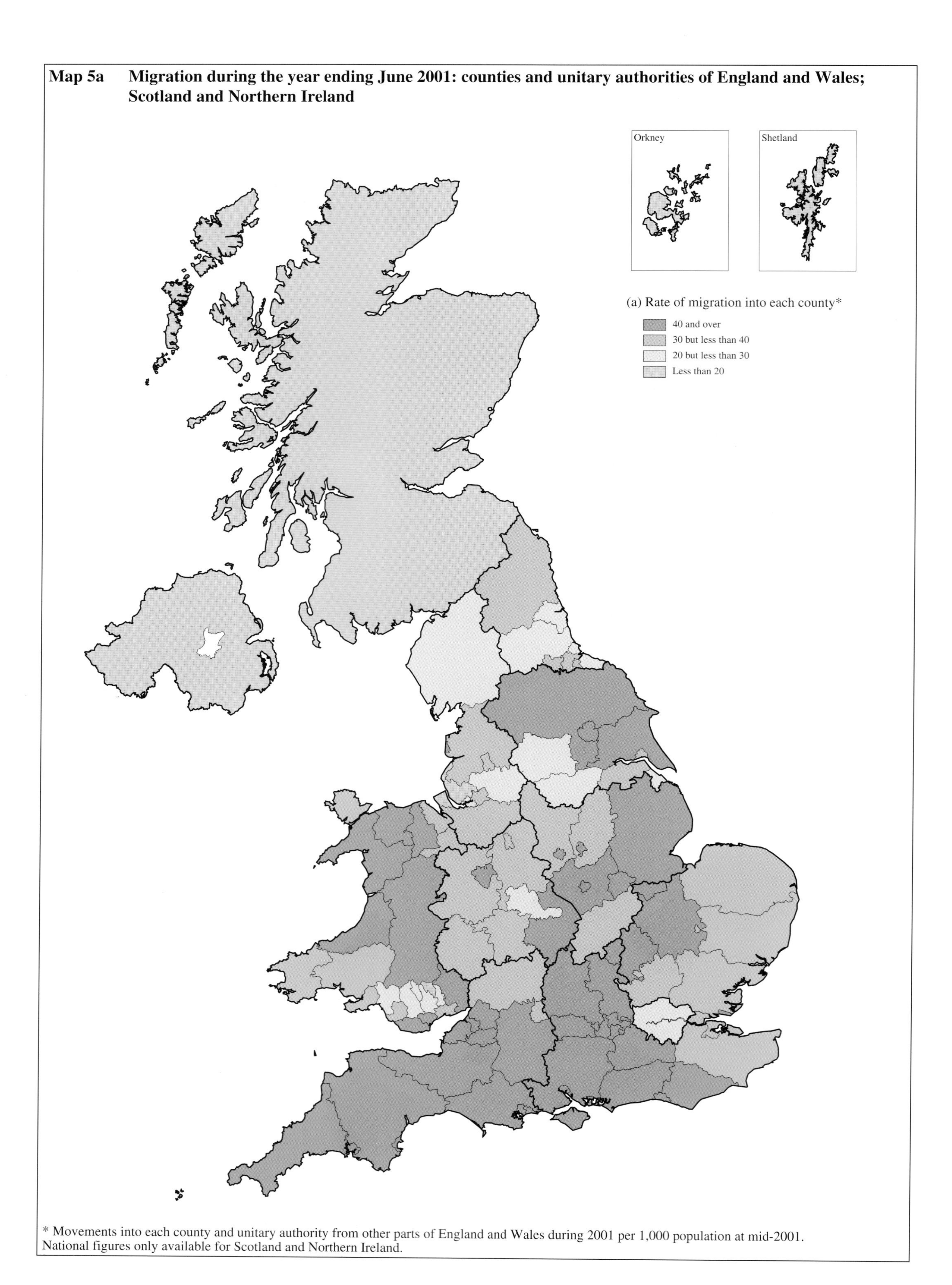

* Movements into each county and unitary authority from other parts of England and Wales during 2001 per 1,000 population at mid-2001. National figures only available for Scotland and Northern Ireland.

Map 5b Migration during the year ending June 2001: counties and unitary authorities of England and Wales; Scotland and Northern Ireland

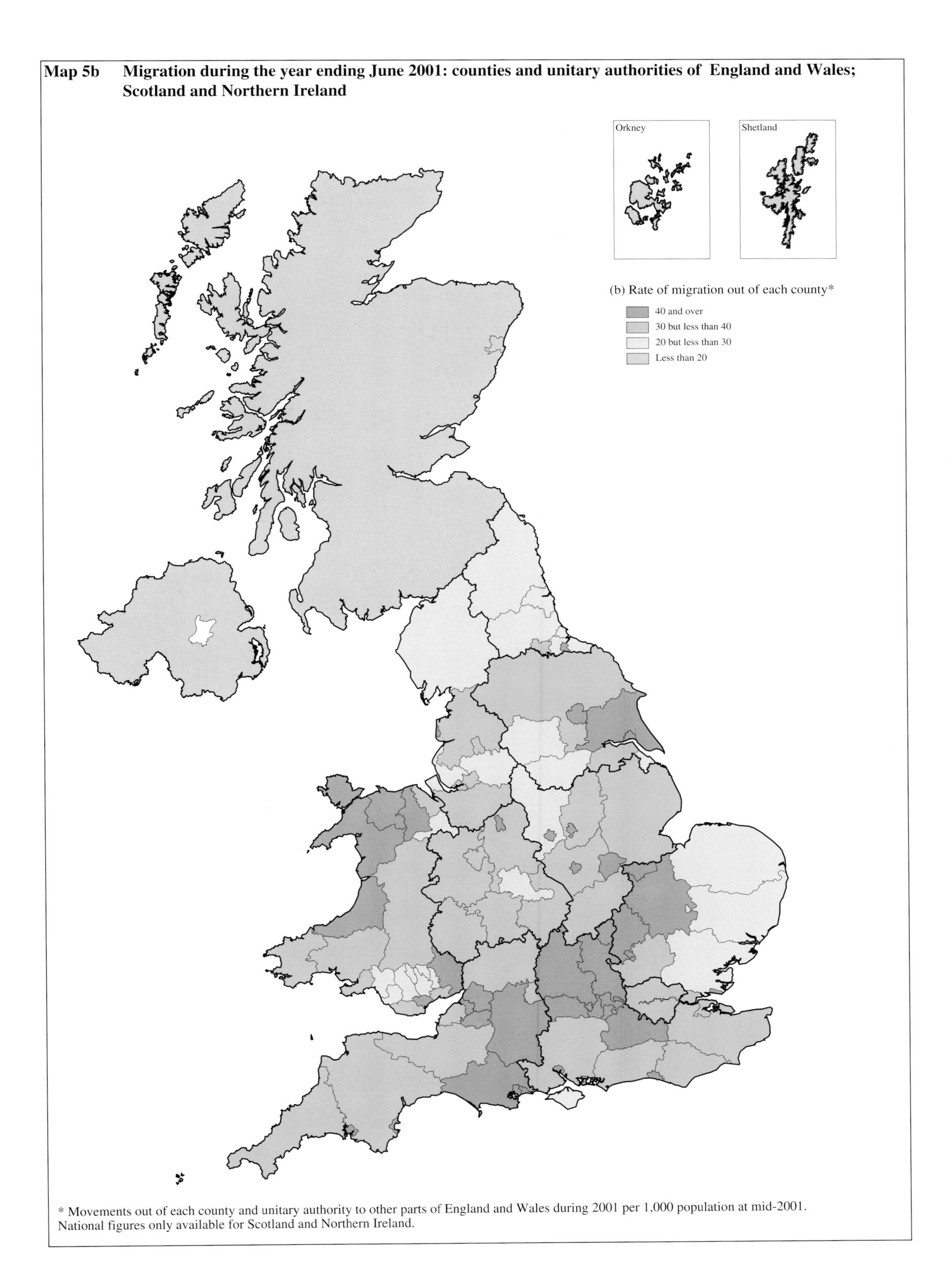

* Movements out of each county and unitary authority to other parts of England and Wales during 2001 per 1,000 population at mid-2001. National figures only available for Scotland and Northern Ireland.

Table 5.1a Movements[1] between health authorities in the United Kingdom during the year ending June 2001: gross and net flows to/from each area by sex and broad age-group

Area	Sex									
	Persons				Males			Females		
	In	Out	Net inflow	Rate[2]	In	Out	Net inflow	In	Out	Net inflow
ENGLAND	**106.0**	**117.0**	**-11.0**	**-0.2**	**51.8**	**56.2**	**-4.4**	**54.2**	**60.8**	**-6.5**
Northern & Yorkshire	**104.3**	**104.8**	**-0.5**	**-0.1**	**50.1**	**50.8**	**-0.6**	**54.2**	**54.1**	**0.1**
Bradford	13.2	15.9	-2.7	-5.8	6.5	7.7	-1.2	6.7	8.2	-1.5
Calderdale and Kirklees	16.1	16.1	0.1	0.1	7.9	7.8	0.1	8.2	8.2	0.0
County Durham and Darlington	16.3	15.8	0.5	0.9	7.9	7.7	0.2	8.4	8.1	0.3
East Riding and Hull	16.7	14.7	2.0	3.6	8.2	7.1	1.1	8.5	7.6	0.9
Gateshead and South Tyneside	7.9	9.2	-1.2	-3.6	4.0	4.6	-0.6	3.9	4.6	-0.7
Leeds	27.2	28.4	-1.2	-1.7	12.9	13.4	-0.5	14.3	15.0	-0.8
Newcastle & North Tyneside	16.9	18.1	-1.2	-2.7	8.2	8.9	-0.8	8.8	9.2	-0.4
North Cumbria	8.3	7.5	0.8	2.5	4.1	3.8	0.3	4.2	3.8	0.5
Northumberland	10.0	9.0	1.0	3.3	4.8	4.3	0.5	5.2	4.7	0.5
North Yorkshire	31.7	28.1	3.6	4.8	14.8	13.2	1.6	17.0	15.0	2.0
Sunderland	5.6	7.0	-1.4	-4.9	2.7	3.6	-0.8	2.9	3.4	-0.5
Tees	10.1	11.6	-1.5	-2.7	5.0	5.8	-0.8	5.1	5.8	-0.6
Wakefield	9.2	8.4	0.8	2.4	4.4	4.1	0.4	4.8	4.3	0.4
Trent	**111.8**	**100.1**	**11.7**	**2.3**	**54.2**	**49.0**	**5.2**	**57.6**	**51.1**	**6.5**
Barnsley	5.5	5.0	0.5	2.4	2.8	2.4	0.3	2.8	2.6	0.2
Doncaster	7.3	7.7	-0.4	-1.3	3.6	3.9	-0.3	3.7	3.8	-0.1
Leicestershire	30.0	28.6	1.4	1.5	14.6	14.0	0.5	15.4	14.6	0.9
Lincolnshire	28.7	21.1	7.6	11.8	13.7	10.0	3.7	15.0	11.1	3.9
North Derbyshire	11.7	10.4	1.3	3.5	5.7	5.2	0.6	6.0	5.3	0.7
North Nottinghamshire	14.1	12.2	1.9	4.9	6.9	6.0	1.0	7.2	6.2	0.9
Nottingham	25.1	25.5	-0.4	-0.6	12.2	12.4	-0.2	12.9	13.0	-0.2
Rotherham	6.5	6.1	0.4	1.6	3.2	3.1	0.1	3.4	3.1	0.3
Sheffield	18.0	20.2	-2.2	-4.3	8.7	9.9	-1.2	9.3	10.3	-1.0
Southern Derbyshire	18.0	16.8	1.1	2.0	8.8	8.3	0.5	9.2	8.6	0.6
South Humber	8.5	8.1	0.4	1.2	4.2	4.0	0.2	4.3	4.1	0.2
Eastern	**144.9**	**125.4**	**19.5**	**3.6**	**70.3**	**60.9**	**9.4**	**74.6**	**64.5**	**10.1**
Bedfordshire	21.0	22.4	-1.4	-2.4	10.3	11.0	-0.7	10.7	11.4	-0.7
Cambridgeshire	33.0	29.7	3.3	4.6	16.2	14.6	1.7	16.7	15.1	1.6
Hertfordshire	41.2	41.4	-0.2	-0.1	19.8	19.9	-0.1	21.4	21.4	0.0
Norfolk	27.7	21.1	6.5	8.2	13.4	10.2	3.2	14.2	10.9	3.3
North Essex	35.3	29.9	5.4	5.9	17.0	14.3	2.7	18.4	15.6	2.7
South Essex	21.6	19.8	1.7	2.5	10.6	9.8	0.8	11.0	10.0	0.9
Suffolk	22.2	18.2	4.1	6.1	10.8	8.9	1.9	11.4	9.2	2.2
London	**163.6**	**232.2**	**-68.6**	**-9.5**	**79.7**	**111.1**	**-31.4**	**83.9**	**121.1**	**-37.2**
Barking and Havering	14.2	14.2	0.0	-0.1	6.7	6.9	-0.2	7.5	7.3	0.1
Barnet, Enfield and Haringey	39.0	47.9	-9.0	-11.1	18.3	22.5	-4.3	20.7	25.4	-4.7
Bexley, Bromley and Greenwich	30.2	31.6	-1.4	-1.9	14.4	15.2	-0.8	15.8	16.4	-0.6
Brent & Harrow	23.0	30.3	-7.3	-15.5	11.0	14.5	-3.5	12.0	15.8	-3.8
Camden & Islington	29.6	33.5	-3.9	-10.3	13.6	14.9	-1.3	16.0	18.6	-2.6
Croydon	15.7	18.1	-2.4	-7.1	7.6	8.8	-1.2	8.1	9.3	-1.2
Ealing, Hammersmith & Hounslow	37.0	47.9	-10.9	-16.1	17.4	22.0	-4.6	19.6	25.9	-6.3
East London and The City	33.9	44.0	-10.2	-15.6	16.4	21.4	-5.0	17.5	22.6	-5.1
Hillingdon	13.2	14.3	-1.0	-4.3	6.4	6.8	-0.4	6.8	7.4	-0.6
Kensington & Chelsea and Westminster	22.9	27.4	-4.5	-13.2	10.7	12.6	-1.9	12.3	14.8	-2.6

Notes: Bold figures exclude moves between health authorities within each health regional office area.
Figures are displayed for health authority boundaries as at 1 April 2001.
These statistics may be revised in light of the 2001 Census results.

1 Based on patient register data and patient re-registration recorded in the NHSCR.
2 Absolute net migration rate per 1,000 resident population calculated using mid-2001 population estimates based on 2001 Census.

United Kingdom by countries and, within England and Wales Health Regional Office areas, and health authorities
thousands

Age 0-15 In	Out	Net inflow	16-29 In	Out	Net inflow	30-59 In	Out	Net inflow	60 and over In	Out	Net inflow	Area
16.4	**21.2**	**-4.7**	**50.0**	**43.6**	**6.4**	**32.9**	**42.1**	**-9.2**	**6.7**	**10.1**	**-3.4**	**ENGLAND**
16.7	**15.3**	**1.4**	**49.5**	**54.0**	**-4.5**	**31.5**	**29.6**	**1.9**	**6.6**	**6.0**	**0.7**	**Northern & Yorkshire**
2.4	3.1	-0.7	5.3	6.3	-1.0	4.5	5.3	-0.7	1.0	1.2	-0.2	Bradford
2.9	2.8	0.2	6.5	6.8	-0.3	5.7	5.4	0.3	1.0	1.1	-0.1	Calderdale and Kirklees
2.9	2.6	0.4	7.1	7.3	-0.2	5.1	4.9	0.2	1.1	0.9	0.1	County Durham and Darlington
3.2	2.1	1.0	6.2	7.4	-1.2	5.7	4.2	1.6	1.7	1.1	0.6	East Riding and Hull
1.5	1.8	-0.2	3.0	3.5	-0.5	2.9	3.4	-0.5	0.5	0.6	0.0	Gateshead and South Tyneside
2.8	3.7	-0.9	17.1	14.2	2.9	6.3	8.6	-2.3	0.9	1.8	-0.9	Leeds
2.1	2.4	-0.3	9.7	9.3	0.4	4.4	5.4	-1.0	0.7	1.0	-0.3	Newcastle & North Tyneside
1.5	1.2	0.3	2.6	3.3	-0.7	3.3	2.4	0.9	0.9	0.6	0.3	North Cumbria
2.1	1.7	0.4	2.5	3.4	-0.9	4.3	3.1	1.1	1.1	0.7	0.4	Northumberland
5.8	4.8	1.0	11.4	12.2	-0.8	11.5	8.9	2.6	3.0	2.2	0.8	North Yorkshire
1.1	1.2	-0.1	2.3	3.1	-0.8	1.9	2.3	-0.4	0.4	0.4	0.0	Sunderland
1.9	1.8	0.2	4.1	5.5	-1.4	3.3	3.5	-0.2	0.7	0.7	0.0	Tees
1.8	1.6	0.3	3.3	3.3	0.0	3.5	3.0	0.4	0.6	0.5	0.0	Wakefield
18.0	**13.9**	**4.1**	**50.9**	**52.4**	**-1.5**	**34.1**	**27.9**	**6.2**	**8.7**	**5.8**	**2.9**	**Trent**
1.2	1.1	0.1	1.8	1.8	0.0	2.2	1.8	0.4	0.4	0.3	0.1	Barnsley
1.5	1.6	0.0	2.4	2.9	-0.5	2.7	2.7	0.0	0.6	0.5	0.0	Doncaster
4.6	3.7	0.9	14.9	15.5	-0.6	8.6	7.8	0.9	1.8	1.6	0.2	Leicestershire
5.8	3.7	2.2	7.6	8.6	-1.0	11.0	6.7	4.3	4.3	2.2	2.1	Lincolnshire
2.5	2.0	0.5	3.4	3.8	-0.5	4.7	3.8	0.9	1.1	0.8	0.3	North Derbyshire
3.2	2.5	0.8	3.9	4.6	-0.7	5.6	4.3	1.3	1.3	0.8	0.5	North Nottinghamshire
2.9	3.5	-0.6	15.3	13.2	2.0	6.0	7.4	-1.4	0.9	1.3	-0.4	Nottingham
1.4	1.2	0.3	2.2	2.3	-0.1	2.4	2.3	0.1	0.5	0.4	0.1	Rotherham
1.9	2.7	-0.8	11.6	10.6	1.0	4.0	5.7	-1.7	0.5	1.2	-0.7	Sheffield
3.3	3.0	0.3	7.0	7.3	-0.3	6.3	5.5	0.9	1.3	1.0	0.3	Southern Derbyshire
2.0	1.6	0.5	2.4	3.3	-0.9	3.2	2.6	0.6	0.8	0.6	0.2	South Humber
24.5	**19.5**	**5.0**	**54.5**	**54.7**	**-0.3**	**52.2**	**42.0**	**10.3**	**13.7**	**9.2**	**4.5**	**Eastern**
3.8	4.0	-0.3	7.9	8.6	-0.8	8.0	8.2	-0.2	1.4	1.6	-0.2	Bedfordshire
5.1	4.5	0.6	14.4	13.4	1.1	10.9	9.8	1.1	2.5	2.0	0.5	Cambridgeshire
6.5	6.3	0.2	16.1	15.5	0.6	15.5	16.1	-0.6	3.1	3.5	-0.4	Hertfordshire
4.9	3.5	1.4	8.4	8.8	-0.4	10.4	6.8	3.6	3.9	2.0	1.9	Norfolk
6.6	5.3	1.3	11.3	11.2	0.1	13.8	10.7	3.1	3.7	2.8	0.9	North Essex
4.2	3.5	0.6	6.8	6.9	0.0	8.1	7.5	0.6	2.4	1.9	0.5	South Essex
4.3	3.2	1.1	6.3	7.2	-0.9	8.8	6.2	2.6	2.9	1.6	1.3	Suffolk
12.6	**41.6**	**-29.0**	**101.4**	**74.8**	**26.7**	**44.0**	**94.5**	**-50.5**	**5.5**	**21.3**	**-15.7**	**London**
3.0	2.8	0.2	4.6	4.5	0.1	5.5	5.5	0.0	1.1	1.4	-0.4	Barking and Havering
4.9	7.6	-2.7	18.9	17.9	1.0	13.4	19.1	-5.7	1.7	3.3	-1.5	Barnet, Enfield and Haringey
4.7	5.3	-0.6	12.3	11.0	1.3	11.4	12.6	-1.2	1.9	2.7	-0.9	Bexley, Bromley and Greenwich
3.2	4.8	-1.6	10.4	11.2	-0.8	8.3	12.1	-3.8	1.1	2.3	-1.2	Brent & Harrow
2.1	4.1	-2.0	17.4	13.8	3.5	9.5	13.8	-4.4	0.7	1.7	-1.0	Camden & Islington
2.5	3.3	-0.7	6.1	5.9	0.2	6.1	7.5	-1.4	1.0	1.4	-0.4	Croydon
3.6	7.5	-3.8	19.7	17.6	2.2	12.6	20.0	-7.4	1.0	2.9	-1.8	Ealing, Hammersmith & Hounslow
3.9	8.6	-4.7	17.1	15.0	2.1	12.1	17.6	-5.5	0.8	2.9	-2.0	East London and The City
2.1	2.6	-0.6	5.9	5.1	0.8	4.5	5.4	-0.9	0.7	1.1	-0.4	Hillingdon
1.8	3.4	-1.6	12.2	10.4	1.7	8.2	11.9	-3.7	0.7	1.7	-1.0	Kensington & Chelsea and Westminster

Table 5.1a - *continued*

Area	Persons In	Persons Out	Persons Net inflow	Persons Rate[2]	Males In	Males Out	Males Net inflow	Females In	Females Out	Females Net inflow
Kingston and Richmond	21.6	22.4	-0.8	-2.5	10.1	10.4	-0.3	11.5	12.0	-0.5
Lambeth, Southwark and Lewisham	46.1	54.6	-8.5	-11.2	21.9	25.3	-3.4	24.1	29.3	-5.2
Merton, Sutton and Wandsworth	37.9	42.9	-5.0	-8.0	16.8	19.6	-2.7	21.1	23.4	-2.3
Redbridge and Waltham Forest	21.9	25.6	-3.7	-8.1	10.6	12.3	-1.8	11.3	13.3	-1.9
South East	**234.1**	**222.9**	**11.2**	**1.3**	**113.4**	**107.9**	**5.5**	**120.6**	**115.0**	**5.6**
Berkshire	34.3	39.3	-5.0	-6.2	16.9	19.1	-2.3	17.5	20.1	-2.7
Buckinghamshire	30.8	30.9	-0.1	-0.2	14.9	15.0	-0.1	15.9	15.9	0.0
East Kent	24.3	20.6	3.8	6.4	11.9	10.1	1.8	12.5	10.5	2.0
East Surrey	22.7	21.9	0.8	1.8	10.8	10.5	0.4	11.9	11.5	0.4
East Sussex, Brighton and Hove	34.0	28.4	5.5	7.5	16.5	13.7	2.8	17.4	14.7	2.7
Isle of Wight, Portsmouth and South East Hampshire	23.3	21.4	1.9	2.9	11.3	10.4	0.8	12.0	10.9	1.1
Northamptonshire	23.6	19.9	3.7	5.8	11.4	9.6	1.9	12.2	10.4	1.8
North and Mid Hampshire	27.9	29.7	-1.8	-3.2	13.4	14.2	-0.9	14.5	15.4	-0.9
Oxfordshire	29.8	31.1	-1.2	-2.0	14.6	15.1	-0.5	15.2	16.0	-0.7
Southampton and South West Hampshire	23.1	21.4	1.6	3.0	11.5	10.7	0.8	11.6	10.7	0.9
West Kent	32.0	31.0	1.0	1.0	15.6	15.1	0.5	16.4	15.9	0.5
West Surrey	31.9	34.3	-2.4	-3.8	15.2	16.5	-1.3	16.7	17.8	-1.1
West Sussex	30.2	26.8	3.4	4.5	14.5	12.9	1.7	15.6	13.9	1.7
South West	**140.2**	**110.4**	**29.8**	**6.0**	**67.8**	**53.0**	**14.8**	**72.4**	**57.5**	**15.0**
Avon	35.4	33.6	1.8	1.9	17.2	16.0	1.2	18.2	17.6	0.6
Cornwall and Isles of Scilly	21.2	15.9	5.3	10.5	10.3	7.7	2.5	10.9	8.2	2.7
Dorset	29.0	23.6	5.3	7.7	14.3	11.5	2.9	14.6	12.2	2.4
Gloucestershire	20.5	18.8	1.7	3.0	9.6	9.0	0.6	10.8	9.7	1.1
North and East Devon	23.7	18.3	5.4	11.1	11.3	8.7	2.7	12.4	9.6	2.7
Somerset	21.6	17.6	4.0	8.0	10.6	8.8	1.8	11.0	8.9	2.2
South and West Devon	25.6	21.2	4.4	7.5	12.7	10.4	2.3	12.9	10.7	2.1
Wiltshire	23.5	21.7	1.7	2.8	11.1	10.4	0.7	12.3	11.3	1.0
West Midlands	**93.5**	**101.0**	**-7.6**	**-1.4**	**44.6**	**49.2**	**-4.6**	**48.9**	**51.9**	**-3.0**
Birmingham	30.3	39.3	-9.0	-9.2	14.5	19.0	-4.4	15.8	20.3	-4.5
Coventry	11.3	12.9	-1.7	-5.6	5.5	6.4	-0.9	5.8	6.5	-0.8
Dudley	8.0	8.6	-0.6	-2.0	3.9	4.3	-0.3	4.0	4.3	-0.3
Herefordshire	7.0	6.2	0.8	4.3	3.3	2.9	0.4	3.6	3.3	0.4
North Staffordshire	11.1	11.8	-0.7	-1.5	5.3	5.7	-0.4	5.8	6.1	-0.2
Sandwell	9.1	10.2	-1.2	-4.1	4.4	4.9	-0.5	4.7	5.4	-0.7
Shropshire	15.6	13.3	2.4	5.4	7.6	6.4	1.2	8.0	6.8	1.2
Solihull	8.6	8.3	0.3	1.5	4.1	3.9	0.2	4.5	4.3	0.1
South Staffordshire	20.6	18.6	2.0	3.4	10.2	9.2	0.9	10.4	9.4	1.1
Walsall	6.9	8.6	-1.7	-6.7	3.4	4.2	-0.9	3.5	4.3	-0.8
Warwickshire	20.6	18.4	2.2	4.3	9.4	8.9	0.5	11.1	9.5	1.6
Wolverhampton	7.4	8.9	-1.5	-6.4	3.5	4.4	-0.9	3.8	4.5	-0.7
Worcestershire	18.5	17.3	1.2	2.2	9.1	8.5	0.6	9.4	8.9	0.6

Notes: Bold figures exclude moves between health authorities within each health regional office area.
Figures are displayed for health authority boundaries as at 1 April 2001.
These statistics may be revised in light of the 2001 Census results.

1 Based on patient register data and patient re-registration recorded in the NHSCR.
2 Absolute net migration rate per 1,000 resident population calculated using mid-2001 population estimates based on 2001 Census.

United Kingdom by countries and, within England and Wales Health Regional Office areas, and health authorities

thousands

0-15 In	0-15 Out	0-15 Net inflow	16-29 In	16-29 Out	16-29 Net inflow	30-59 In	30-59 Out	30-59 Net inflow	60 and over In	60 and over Out	60 and over Net inflow	Area
2.1	3.0	-0.9	10.8	8.7	2.1	7.8	9.1	-1.3	0.9	1.6	-0.7	Kingston and Richmond
3.7	9.0	-5.4	25.6	19.6	6.1	15.7	22.8	-7.1	1.1	3.2	-2.1	Lambeth, Southwark and Lewisham
3.2	6.3	-3.1	21.3	15.4	5.9	12.2	18.5	-6.3	1.3	2.7	-1.4	Merton, Sutton and Wandsworth
3.7	5.1	-1.4	8.7	8.3	0.4	8.3	10.1	-1.9	1.3	2.1	-0.8	Redbridge and Waltham Forest
38.5	**33.9**	**4.5**	**93.5**	**96.0**	**-2.5**	**83.6**	**74.9**	**8.7**	**18.5**	**18.1**	**0.4**	**South East**
5.1	6.5	-1.4	16.1	15.1	0.9	11.4	14.8	-3.4	1.7	2.9	-1.1	Berkshire
5.6	5.2	0.4	11.0	11.5	-0.6	12.1	11.8	0.3	2.1	2.4	-0.2	Buckinghamshire
4.8	3.5	1.3	7.7	9.1	-1.3	8.9	6.0	2.8	3.0	2.0	1.0	East Kent
4.1	3.4	0.7	7.3	7.8	-0.6	9.5	8.5	1.0	1.8	2.2	-0.4	East Surrey
4.5	3.8	0.7	13.1	12.0	1.1	12.5	9.6	2.8	3.8	3.0	0.8	East Sussex, Brighton and Hove
4.2	3.8	0.4	8.2	8.7	-0.6	8.4	7.0	1.4	2.5	1.8	0.6	Isle of Wight, Portsmouth and South East Hampshire
4.9	3.5	1.4	7.8	8.2	-0.3	9.1	6.8	2.3	1.8	1.5	0.3	Northamptonshire
5.5	5.6	-0.2	10.1	10.6	-0.6	10.4	11.1	-0.7	2.0	2.4	-0.4	North and Mid Hampshire
4.5	4.6	-0.1	14.2	14.5	-0.3	9.5	10.1	-0.6	1.6	1.9	-0.2	Oxfordshire
3.1	2.8	0.3	11.1	10.5	0.6	7.0	6.5	0.6	1.9	1.7	0.2	Southampton and South West Hampshire
6.3	5.5	0.9	10.7	11.4	-0.7	12.3	11.2	1.1	2.6	2.9	-0.3	West Kent
4.5	5.1	-0.6	13.9	13.5	0.3	11.4	12.7	-1.3	2.1	2.9	-0.9	West Surrey
5.1	4.3	0.8	9.2	9.9	-0.7	11.8	9.5	2.3	4.0	3.1	0.9	West Sussex
23.5	**15.6**	**7.9**	**51.1**	**51.8**	**-0.7**	**50.6**	**33.8**	**16.8**	**15.0**	**9.3**	**5.7**	**South West**
3.9	4.4	-0.5	19.8	16.1	3.7	9.9	10.8	-0.9	1.8	2.3	-0.5	Avon
4.0	2.5	1.6	5.4	6.5	-1.1	9.0	5.3	3.7	2.9	1.7	1.2	Cornwall and Isles of Scilly
4.7	3.3	1.4	9.6	10.1	-0.5	10.5	7.4	3.1	4.1	2.9	1.3	Dorset
3.4	2.9	0.5	7.8	8.0	-0.2	7.4	6.3	1.1	1.9	1.6	0.3	Gloucestershire
4.0	2.5	1.4	7.6	8.0	-0.4	8.9	5.7	3.2	3.3	2.0	1.2	North and East Devon
4.6	2.7	1.8	5.8	7.2	-1.4	8.5	5.9	2.6	2.8	1.8	1.0	Somerset
4.4	3.2	1.2	8.8	9.2	-0.4	9.5	6.6	2.8	3.0	2.1	0.8	South and West Devon
4.8	4.3	0.5	7.3	7.8	-0.5	9.3	8.0	1.3	2.1	1.7	0.4	Wiltshire
15.3	**15.0**	**0.3**	**43.0**	**49.4**	**-6.3**	**28.8**	**29.8**	**-1.0**	**6.3**	**6.8**	**-0.5**	**West Midlands**
4.3	6.7	-2.4	16.4	16.8	-0.4	8.2	13.0	-4.7	1.3	2.8	-1.5	Birmingham
1.4	1.9	-0.5	7.0	6.9	0.1	2.5	3.5	-1.0	0.4	0.7	-0.4	Coventry
1.7	1.6	0.1	2.6	3.0	-0.4	3.0	3.2	-0.2	0.6	0.7	-0.1	Dudley
1.4	1.0	0.3	1.8	2.4	-0.6	2.9	2.1	0.7	1.0	0.7	0.3	Herefordshire
1.7	1.8	-0.1	5.5	6.1	-0.5	3.2	3.2	-0.1	0.7	0.7	0.0	North Staffordshire
2.1	2.3	-0.2	3.3	3.4	-0.1	3.2	3.8	-0.6	0.5	0.7	-0.3	Sandwell
3.5	2.3	1.1	4.7	5.6	-1.0	5.9	4.2	1.7	1.6	1.1	0.5	Shropshire
1.8	1.3	0.5	2.6	3.0	-0.5	3.4	3.1	0.4	0.7	0.9	-0.2	Solihull
4.0	3.4	0.7	6.6	7.2	-0.6	7.9	6.6	1.2	2.1	1.4	0.7	South Staffordshire
1.6	1.8	-0.2	2.3	2.9	-0.6	2.5	3.1	-0.6	0.4	0.8	-0.3	Walsall
3.6	2.9	0.6	7.3	7.6	-0.2	8.2	6.5	1.6	1.5	1.4	0.1	Warwickshire
1.4	1.8	-0.3	3.0	3.6	-0.6	2.5	3.0	-0.5	0.4	0.5	-0.1	Wolverhampton
3.5	2.9	0.6	5.7	6.8	-1.1	7.3	6.3	1.0	2.1	1.4	0.7	Worcestershire

Table 5.1a - *continued*

Area	Sex									
	Persons				Males			Females		
	In	Out	Net inflow	Rate[2]	In	Out	Net inflow	In	Out	Net inflow
North West	**101.4**	**107.8**	**-6.4**	**-1.0**	**49.4**	**52.2**	**-2.8**	**52.0**	**55.6**	**-3.6**
Bury and Rochdale	11.9	12.9	-1.0	-2.5	6.0	6.5	-0.5	6.0	6.4	-0.4
East Lancashire	12.2	13.5	-1.2	-2.4	6.2	6.8	-0.6	6.1	6.7	-0.6
Liverpool	15.3	18.4	-3.2	-7.2	7.2	8.6	-1.4	8.1	9.8	-1.7
Manchester	27.0	30.0	-3.0	-7.7	13.4	14.3	-0.9	13.6	15.7	-2.1
Morecambe Bay	12.3	11.2	1.1	3.6	5.9	5.3	0.6	6.4	5.9	0.6
North Cheshire	8.8	9.2	-0.3	-1.1	4.4	4.6	-0.1	4.4	4.6	-0.2
North-West Lancashire	18.9	16.9	2.0	4.3	9.2	8.3	0.8	9.7	8.6	1.1
St Helens and Knowsley	8.2	8.8	-0.6	-1.8	4.1	4.4	-0.3	4.2	4.5	-0.3
Salford and Trafford	16.3	18.5	-2.2	-5.3	8.2	9.2	-1.0	8.1	9.3	-1.2
Sefton	8.0	7.6	0.4	1.3	3.8	3.8	0.0	4.2	3.8	0.4
South Cheshire	25.0	22.9	2.1	3.1	12.0	11.1	1.0	13.0	11.9	1.1
South Lancashire	11.1	10.0	1.2	3.7	5.3	4.7	0.6	5.8	5.2	0.6
Stockport	9.8	10.4	-0.6	-2.0	4.7	5.0	-0.3	5.1	5.4	-0.3
West Pennine	11.2	12.0	-0.8	-1.8	5.5	6.0	-0.5	5.7	6.0	-0.3
Wigan and Bolton	13.4	13.4	0.1	0.1	6.6	6.6	0.0	6.9	6.8	0.1
	0.0	0.0	0.0		0.0	0.0	0.0	0.0	0.0	0.0
Wirral	7.1	7.3	-0.2	-0.6	3.5	3.5	0.0	3.6	3.7	-0.1
WALES	**59.3**	**51.8**	**7.5**	**2.6**	**28.6**	**25.1**	**3.5**	**30.7**	**26.6**	**4.1**
North Wales	19.6	16.5	3.1	4.7	9.5	8.0	1.5	10.1	8.4	1.7
Dyfed Powys	18.2	15.6	2.6	5.3	8.8	7.6	1.2	9.4	8.0	1.3
Morgannwg	12.0	10.8	1.3	2.6	5.8	5.2	0.6	6.2	5.6	0.7
Bro Taf	19.9	20.1	-0.3	-0.4	9.3	9.4	-0.1	10.6	10.7	-0.1
Gwent	11.6	10.9	0.8	1.4	5.6	5.4	0.2	6.0	5.5	0.6
SCOTLAND	**54.9**	**51.5**	**3.4**	**0.7**	**26.2**	**25.2**	**1.0**	**28.7**	**26.3**	**2.4**
NORTHERN IRELAND	**11.5**	**11.5**	**0.0**	**0.0**	**5.1**	**5.1**	**0.0**	**6.4**	**6.4**	**0.0**

Notes: Bold figures exclude moves between health authorities within each health regional office area.
Figures are displayed for health authority boundaries as at 1 April 2001.
These statistics may be revised in light of the 2001 Census results.

1 Based on patient register data and patient re-registration recorded in the NHSCR.
2 Absolute net migration rate per 1,000 resident population calculated using mid-2001 population estimates based on 2001 Census.

United Kingdom by countries and, within England and Wales Health Regional Office areas, and health authorities
thousands

Age												Area
0-15			16-29			30-59			60 and over			
In	Out	Net inflow	In	Out	Net inflow	In	Out	Net inflow	In	Out	Net inflow	
16.2	**15.3**	**0.9**	**48.2**	**52.6**	**-4.4**	**30.7**	**32.3**	**-1.6**	**6.3**	**7.6**	**-1.3**	**North West**
2.5	2.5	0.0	4.1	4.7	-0.5	4.5	4.7	-0.2	0.9	1.0	-0.2	Bury and Rochdale
2.7	2.8	-0.1	4.1	5.1	-1.0	4.5	4.4	0.1	0.9	1.1	-0.2	East Lancashire
2.1	2.9	-0.8	8.8	8.8	0.0	3.8	5.6	-1.8	0.6	1.2	-0.6	Liverpool
3.3	5.0	-1.7	16.0	14.0	2.0	6.8	9.5	-2.7	0.9	1.5	-0.7	Manchester
2.0	1.5	0.6	5.2	5.9	-0.7	3.9	3.0	0.9	1.2	0.8	0.4	Morecambe Bay
1.8	1.7	0.1	3.0	3.3	-0.3	3.4	3.5	-0.1	0.6	0.7	0.0	North Cheshire
3.7	2.9	0.8	6.5	7.0	-0.5	6.5	5.4	1.1	2.2	1.6	0.7	North-West Lancashire
1.9	2.0	0.0	2.6	3.0	-0.4	3.1	3.2	-0.1	0.5	0.7	-0.1	St Helens and Knowsley
2.6	3.0	-0.4	7.3	7.5	-0.2	5.6	6.6	-1.0	0.8	1.4	-0.6	Salford and Trafford
1.7	1.2	0.5	2.3	2.9	-0.5	3.1	2.7	0.4	0.9	0.8	0.1	Sefton
4.4	3.4	0.9	9.2	9.7	-0.5	9.5	8.1	1.4	2.0	1.7	0.3	South Cheshire
2.3	1.6	0.6	3.5	4.0	-0.4	4.3	3.6	0.7	1.0	0.8	0.3	South Lancashire
1.8	1.7	0.1	3.5	3.7	-0.3	3.8	4.0	-0.1	0.7	1.0	-0.2	Stockport
2.3	2.3	0.0	3.9	4.1	-0.2	4.2	4.5	-0.3	0.8	1.0	-0.3	West Pennine
2.8	2.5	0.2	4.8	5.0	-0.2	4.9	4.9	0.1	0.9	0.9	0.0	Wigan and Bolton
0.0	0.0	0.0	0.0	0.0	0.0	0.0	0.0	0.0	0.0	0.0	0.0	
1.3	1.1	0.2	2.4	3.0	-0.6	2.6	2.5	0.2	0.7	0.7	0.0	Wirral
9.9	**7.3**	**2.6**	**23.0**	**26.1**	**-3.1**	**20.4**	**14.4**	**6.0**	**6.0**	**4.0**	**2.0**	**WALES**
3.6	2.7	0.9	6.1	7.2	-1.1	7.4	5.0	2.5	2.6	1.7	0.9	North Wales
3.2	2.3	0.9	5.8	7.6	-1.8	6.8	4.3	2.5	2.4	1.4	1.0	Dyfed Powys
2.1	1.5	0.6	5.2	5.6	-0.3	3.8	3.0	0.8	0.9	0.7	0.2	Morgannwg
2.5	2.8	-0.3	11.4	10.6	0.8	5.1	5.7	-0.6	0.9	1.1	-0.1	Bro Taf
2.4	1.9	0.5	3.8	4.5	-0.7	4.5	3.6	0.8	0.9	0.9	0.1	Gwent
10.0	**8.5**	**1.4**	**21.1**	**22.8**	**-1.8**	**20.2**	**17.5**	**2.7**	**3.7**	**2.7**	**1.0**	**SCOTLAND**
2.7	**2.2**	**0.4**	**4.4**	**6.2**	**-1.8**	**3.9**	**2.7**	**1.2**	**0.6**	**0.4**	**0.2**	**NORTHERN IRELAND**

Table 5.1b Movements[1] between local authorities in the United Kingdom during the year ending June 2001: gross and net flows to/from each area by sex and by broad age-group

Area	Sex									
	Persons				Males			Females		
	In	Out	Net inflow	Rate[2]	In	Out	Net inflow	In	Out	Net inflow
ENGLAND	**106.0**	**117.0**	**-11.0**	**-0.2**	**51.8**	**56.2**	**-4.4**	**54.2**	**60.8**	**-6.5**
NORTH EAST	**39.5**	**43.2**	**-3.7**	**-1.5**	**19.1**	**21.4**	**-2.3**	**20.5**	**21.9**	**-1.4**
Darlington UA	3.5	3.4	0.1	1.5	1.7	1.7	0.0	1.8	1.7	0.1
Hartlepool UA	2.1	1.9	0.2	1.7	1.0	1.0	0.1	1.1	1.0	0.1
Middlesbrough UA	4.6	6.2	-1.6	-11.9	2.4	3.1	-0.8	2.3	3.1	-0.8
Redcar and Cleveland UA	3.6	4.1	-0.5	-3.4	1.8	2.0	-0.2	1.8	2.0	-0.2
Stockton-on-Tees UA	5.6	5.2	0.5	2.6	2.7	2.6	0.1	2.9	2.6	0.3
Durham	**14.3**	**13.9**	**0.4**	**0.7**	**6.9**	**6.8**	**0.1**	**7.4**	**7.1**	**0.2**
Chester-le-Street	2.1	2.2	-0.1	-2.4	1.0	1.1	0.0	1.1	1.2	-0.1
Derwentside	3.0	2.5	0.5	5.5	1.5	1.2	0.2	1.5	1.3	0.2
Durham	6.4	6.1	0.3	3.4	3.0	2.9	0.1	3.4	3.2	0.2
Easington	1.9	2.3	-0.3	-3.7	1.0	1.2	-0.2	1.0	1.1	-0.2
Sedgefield	3.1	3.1	0.0	0.4	1.5	1.5	0.0	1.6	1.5	0.0
Teesdale	1.2	1.1	0.0	1.6	0.6	0.5	0.0	0.6	0.6	0.0
Wear Valley	2.2	2.2	0.0	-0.1	1.1	1.1	0.0	1.1	1.1	0.0
Northumberland	**10.0**	**9.0**	**1.0**	**3.3**	**4.8**	**4.3**	**0.5**	**5.2**	**4.7**	**0.5**
Alnwick	1.7	1.5	0.2	7.8	0.8	0.7	0.1	0.9	0.8	0.1
Berwick-upon-Tweed	1.3	1.1	0.1	4.8	0.6	0.5	0.1	0.6	0.6	0.0
Blyth Valley	3.0	2.9	0.1	1.4	1.5	1.4	0.1	1.5	1.5	0.0
Castle Morpeth	2.7	2.6	0.0	0.5	1.3	1.3	0.1	1.4	1.4	0.0
Tynedale	2.7	2.2	0.5	8.0	1.3	1.1	0.2	1.4	1.1	0.3
Wansbeck	2.0	1.9	0.0	0.6	0.9	1.0	0.0	1.0	1.0	0.1
Tyne and Wear (Met County)	**23.5**	**27.3**	**-3.8**	**-3.5**	**11.3**	**13.5**	**-2.2**	**12.2**	**13.8**	**-1.6**
Gateshead	5.7	6.4	-0.7	-3.6	2.9	3.1	-0.3	2.8	3.2	-0.4
Newcastle upon Tyne	14.0	16.3	-2.2	-8.6	6.8	8.0	-1.2	7.2	8.2	-1.0
North Tyneside	7.4	6.4	1.0	5.4	3.6	3.2	0.5	3.8	3.2	0.6
South Tyneside	3.0	3.6	-0.5	-3.6	1.5	1.8	-0.3	1.5	1.7	-0.3
Sunderland	5.6	7.0	-1.4	-4.9	2.7	3.6	-0.8	2.9	3.4	-0.5
NORTH WEST	**105.8**	**111.4**	**-5.7**	**-0.8**	**51.5**	**54.0**	**-2.5**	**54.3**	**57.4**	**-3.2**
Blackburn with Darwen UA	4.5	5.4	-0.8	-6.0	2.3	2.7	-0.5	2.3	2.6	-0.4
Blackpool UA	8.8	8.4	0.4	2.7	4.5	4.3	0.2	4.3	4.2	0.1
Halton UA	3.3	3.7	-0.5	-4.0	1.6	1.9	-0.3	1.6	1.8	-0.2
Warrington UA	6.4	6.3	0.1	0.7	3.2	3.1	0.1	3.2	3.2	0.0
Cheshire	**25.0**	**22.9**	**2.1**	**3.1**	**12.0**	**11.1**	**1.0**	**13.0**	**11.9**	**1.1**
Chester	6.7	6.9	-0.2	-1.6	3.1	3.2	-0.1	3.6	3.7	-0.1
Congleton	4.9	4.4	0.4	4.9	2.4	2.2	0.2	2.5	2.2	0.3
Crewe and Nantwich	5.0	3.7	1.3	12.0	2.5	1.8	0.7	2.5	1.8	0.7
Ellesmere Port and Neston	2.6	2.9	-0.3	-4.0	1.3	1.5	-0.2	1.3	1.5	-0.1
Macclesfield	6.5	6.7	-0.3	-1.7	3.1	3.3	-0.2	3.4	3.5	-0.1
Vale Royal	5.5	4.5	1.1	8.6	2.7	2.2	0.6	2.8	2.3	0.5
Cumbria	**13.2**	**12.0**	**1.1**	**2.3**	**6.5**	**6.0**	**0.5**	**6.7**	**6.0**	**0.6**
Allerdale	3.2	2.8	0.4	4.0	1.6	1.4	0.1	1.6	1.4	0.2
Barrow-in-Furness	1.6	2.0	-0.4	-4.9	0.8	1.0	-0.2	0.8	1.0	-0.2
Carlisle	3.6	3.5	0.1	0.8	1.7	1.7	0.0	1.9	1.8	0.1
Copeland	1.7	2.0	-0.3	-4.1	0.9	1.0	-0.2	0.8	0.9	-0.1
Eden	2.5	1.9	0.6	11.9	1.3	1.0	0.3	1.2	0.9	0.3
South Lakeland	5.1	4.4	0.7	6.8	2.5	2.2	0.3	2.6	2.3	0.4
Greater Manchester (Met County)	**52.0**	**59.6**	**-7.6**	**-3.1**	**25.6**	**28.9**	**-3.3**	**26.4**	**30.7**	**-4.3**
Bolton	8.3	7.9	0.4	1.4	4.2	3.9	0.2	4.1	4.0	0.2
Bury	6.8	7.1	-0.3	-1.7	3.4	3.6	-0.2	3.4	3.6	-0.1
Manchester	27.0	30.0	-3.0	-7.7	13.4	14.3	-0.9	13.6	15.7	-2.1
Oldham	5.5	6.4	-0.8	-3.8	2.7	3.2	-0.5	2.8	3.2	-0.4
Rochdale	6.3	7.0	-0.7	-3.2	3.1	3.5	-0.3	3.2	3.5	-0.3
Salford	8.8	10.7	-1.8	-8.5	4.6	5.5	-0.9	4.2	5.2	-1.0
Stockport	9.8	10.4	-0.6	-2.0	4.7	5.0	-0.3	5.1	5.4	-0.3
Tameside	6.7	6.7	0.0	-0.1	3.3	3.3	-0.1	3.4	3.4	0.0
Trafford	8.8	9.2	-0.4	-1.9	4.2	4.4	-0.2	4.6	4.8	-0.2
Wigan	7.3	7.6	-0.3	-1.1	3.5	3.8	-0.3	3.8	3.8	-0.1

Note: Bold figures exclude moves between local authorities within each county and moves between counties and unitary authorities within each Government Office Region.
These statistics may be revised in light of the 2001 Census results.

1 Based on patient register data and patient re-registration recorded in the NHSCR.
2 Absolute net migration rate per 1,000 resident population calculated using mid-2001 population estimates based on 2001 Census.

United Kingdom by countries and, within England and Wales Government Office Regions, counties/unitary authorities, districts and London boroughs

thousands

Age												Area
0-15			16-29			30-59			60 and over			
In	Out	Net inflow	In	Out	Net inflow	In	Out	Net inflow	In	Out	Net inflow	
16.4	**21.2**	**-4.7**	**50.0**	**43.6**	**6.4**	**32.9**	**42.1**	**-9.2**	**6.7**	**10.1**	**-3.4**	**ENGLAND**
6.3	**6.0**	**0.3**	**19.4**	**22.8**	**-3.4**	**11.4**	**12.2**	**-0.8**	**2.4**	**2.3**	**0.1**	**NORTH EAST**
0.7	0.7	0.0	1.1	1.1	0.0	1.4	1.3	0.1	0.3	0.3	0.0	**Darlington UA**
0.5	0.4	0.1	0.7	0.8	-0.1	0.8	0.7	0.1	0.2	0.1	0.0	**Hartlepool UA**
0.7	1.1	-0.4	2.2	2.8	-0.5	1.4	2.0	-0.6	0.3	0.4	-0.1	**Middlesbrough UA**
0.9	0.7	0.1	1.0	1.5	-0.6	1.4	1.4	0.0	0.4	0.4	0.0	**Redcar and Cleveland UA**
1.2	0.9	0.3	2.0	2.2	-0.2	2.0	1.7	0.3	0.4	0.3	0.0	**Stockton-on-Tees UA**
2.6	**2.3**	**0.3**	**6.4**	**6.6**	**-0.2**	**4.4**	**4.2**	**0.2**	**0.9**	**0.8**	**0.1**	**Durham**
0.4	0.5	-0.1	0.6	0.7	-0.1	0.9	0.9	0.0	0.2	0.2	0.0	Chester-le-Street
0.8	0.6	0.2	0.9	0.8	0.1	1.1	1.0	0.1	0.3	0.2	0.1	Derwentside
0.7	0.7	0.0	4.0	3.7	0.4	1.4	1.5	-0.1	0.2	0.3	0.0	Durham
0.5	0.5	0.0	0.6	0.8	-0.1	0.7	0.8	-0.1	0.1	0.2	-0.1	Easington
0.8	0.7	0.1	0.9	1.1	-0.1	1.2	1.1	0.1	0.3	0.2	0.0	Sedgefield
0.3	0.2	0.1	0.3	0.4	-0.1	0.5	0.4	0.1	0.1	0.1	0.0	Teesdale
0.5	0.5	0.1	0.6	0.7	-0.2	0.9	0.8	0.1	0.2	0.2	0.0	Wear Valley
2.1	**1.7**	**0.4**	**2.5**	**3.4**	**-0.9**	**4.3**	**3.1**	**1.1**	**1.1**	**0.7**	**0.4**	**Northumberland**
0.4	0.4	0.0	0.3	0.5	-0.1	0.8	0.5	0.2	0.2	0.1	0.1	Alnwick
0.2	0.2	0.0	0.3	0.4	-0.1	0.6	0.4	0.2	0.2	0.2	0.1	Berwick-upon-Tweed
0.6	0.6	0.0	0.8	1.0	-0.1	1.2	1.1	0.1	0.3	0.2	0.1	Blyth Valley
0.6	0.5	0.1	0.6	0.9	-0.3	1.2	1.0	0.2	0.2	0.2	0.0	Castle Morpeth
0.6	0.4	0.2	0.6	0.8	-0.2	1.2	0.8	0.4	0.3	0.2	0.1	Tynedale
0.4	0.4	0.0	0.6	0.6	0.0	0.8	0.7	0.1	0.2	0.2	0.0	Wansbeck
3.3	**4.0**	**-0.7**	**12.5**	**13.4**	**-0.9**	**6.5**	**8.4**	**-1.9**	**1.2**	**1.5**	**-0.3**	**Tyne and Wear (Met County)**
1.1	1.2	-0.2	2.2	2.4	-0.2	2.1	2.4	-0.3	0.4	0.4	0.0	Gateshead
1.5	2.1	-0.6	8.7	8.5	0.3	3.3	4.9	-1.5	0.5	0.8	-0.3	Newcastle upon Tyne
1.4	1.1	0.3	2.4	2.3	0.2	2.9	2.4	0.5	0.6	0.6	0.0	North Tyneside
0.6	0.7	-0.1	1.0	1.3	-0.3	1.1	1.3	-0.1	0.2	0.2	0.0	South Tyneside
1.1	1.2	-0.1	2.3	3.1	-0.8	1.9	2.3	-0.4	0.4	0.4	0.0	Sunderland
17.0	**15.9**	**1.1**	**49.3**	**54.3**	**-5.0**	**32.6**	**33.4**	**-0.8**	**6.8**	**7.8**	**-1.0**	**NORTH WEST**
1.0	1.2	-0.2	1.6	1.8	-0.2	1.6	1.9	-0.2	0.3	0.5	-0.2	**Blackburn with Darwen UA**
1.9	1.7	0.2	2.5	2.5	0.0	3.5	3.3	0.2	1.0	1.0	0.0	**Blackpool UA**
0.7	0.7	0.0	1.0	1.3	-0.2	1.2	1.4	-0.2	0.2	0.4	-0.1	**Halton UA**
1.2	1.1	0.1	2.3	2.4	-0.1	2.5	2.4	0.0	0.4	0.4	0.1	**Warrington UA**
4.4	**3.4**	**0.9**	**9.2**	**9.7**	**-0.5**	**9.5**	**8.1**	**1.4**	**2.0**	**1.7**	**0.3**	**Cheshire**
0.9	1.0	-0.1	3.2	3.0	0.1	2.2	2.4	-0.2	0.4	0.5	-0.1	Chester
0.9	0.7	0.2	1.6	1.8	-0.2	1.9	1.6	0.3	0.4	0.3	0.1	Congleton
1.0	0.6	0.4	1.6	1.5	0.0	2.0	1.2	0.7	0.4	0.2	0.2	Crewe and Nantwich
0.5	0.5	0.0	0.9	1.0	-0.1	0.9	1.1	-0.2	0.2	0.3	0.0	Ellesmere Port and Neston
1.1	1.1	0.1	2.0	2.3	-0.3	2.7	2.7	0.0	0.6	0.6	-0.1	Macclesfield
1.1	0.8	0.3	1.5	1.6	-0.1	2.4	1.8	0.7	0.5	0.4	0.1	Vale Royal
2.4	**1.8**	**0.6**	**4.0**	**5.3**	**-1.3**	**5.2**	**3.8**	**1.4**	**1.5**	**1.0**	**0.4**	**Cumbria**
0.7	0.5	0.2	0.8	1.1	-0.2	1.4	1.0	0.4	0.3	0.3	0.0	Allerdale
0.3	0.4	0.0	0.5	0.7	-0.2	0.6	0.7	-0.1	0.2	0.2	0.0	Barrow-in-Furness
0.6	0.6	0.0	1.3	1.4	-0.1	1.3	1.2	0.0	0.4	0.2	0.1	Carlisle
0.3	0.4	-0.1	0.5	0.7	-0.2	0.7	0.7	0.0	0.2	0.2	0.0	Copeland
0.5	0.3	0.2	0.7	0.8	-0.1	1.1	0.6	0.5	0.3	0.2	0.1	Eden
1.0	0.6	0.4	1.4	1.8	-0.5	2.1	1.5	0.6	0.7	0.5	0.2	South Lakeland
7.3	**9.3**	**-1.9**	**27.4**	**26.7**	**0.7**	**14.9**	**19.2**	**-4.3**	**2.4**	**4.4**	**-2.0**	**Greater Manchester (Met County)**
1.8	1.5	0.3	3.0	3.0	0.0	3.0	2.9	0.1	0.6	0.6	0.0	Bolton
1.4	1.4	0.0	2.3	2.5	-0.2	2.6	2.7	-0.1	0.5	0.6	-0.1	Bury
3.3	5.0	-1.7	16.0	14.0	2.0	6.8	9.5	-2.7	0.9	1.5	-0.7	Manchester
1.2	1.3	-0.1	2.0	2.2	-0.2	2.0	2.4	-0.4	0.4	0.6	-0.2	Oldham
1.4	1.5	-0.1	2.1	2.4	-0.3	2.4	2.5	-0.1	0.4	0.6	-0.1	Rochdale
1.2	1.9	-0.6	4.4	4.4	0.0	2.7	3.7	-1.0	0.5	0.7	-0.3	Salford
1.8	1.7	0.1	3.5	3.7	-0.3	3.8	4.0	-0.1	0.7	1.0	-0.2	Stockport
1.4	1.4	0.0	2.2	2.1	0.1	2.6	2.6	0.0	0.5	0.6	-0.1	Tameside
1.6	1.4	0.2	3.3	3.5	-0.2	3.4	3.4	0.0	0.5	0.8	-0.3	Trafford
1.5	1.5	0.0	2.4	2.6	-0.2	2.8	2.9	0.0	0.6	0.6	0.0	Wigan

Table 5.1b - *continued*

Area	Sex									
	Persons				Males			Females		
	In	Out	Net inflow	Rate[2]	In	Out	Net inflow	In	Out	Net inflow
Lancashire	**37.6**	**34.5**	**3.1**	**2.7**	**17.9**	**16.5**	**1.4**	**19.7**	**18.0**	**1.7**
Burnley	3.2	3.8	-0.6	-6.9	1.6	1.8	-0.2	1.6	1.9	-0.4
Chorley	4.4	4.0	0.4	4.0	2.1	1.9	0.2	2.3	2.1	0.2
Fylde	5.0	4.1	0.9	12.7	2.4	1.9	0.4	2.6	2.1	0.5
Hyndburn	3.2	3.3	-0.1	-1.4	1.6	1.6	-0.1	1.6	1.6	-0.1
Lancaster	7.8	7.0	0.8	5.7	3.6	3.2	0.4	4.2	3.8	0.4
Pendle	3.2	3.4	-0.2	-2.2	1.6	1.7	-0.1	1.6	1.7	-0.1
Preston	6.3	7.0	-0.7	-5.5	2.9	3.3	-0.5	3.4	3.7	-0.3
Ribble Valley	3.0	2.5	0.6	10.6	1.5	1.2	0.2	1.6	1.2	0.4
Rossendale	2.8	2.9	-0.1	-0.8	1.4	1.4	0.0	1.4	1.5	-0.1
South Ribble	4.3	4.1	0.2	2.4	2.1	2.0	0.1	2.2	2.1	0.1
West Lancashire	4.6	4.1	0.5	4.7	2.1	1.9	0.3	2.4	2.2	0.3
Wyre	6.3	4.9	1.4	12.9	3.0	2.4	0.6	3.2	2.5	0.7
Merseyside (Met County)	**25.5**	**29.1**	**-3.6**	**-2.6**	**12.2**	**13.9**	**-1.7**	**13.3**	**15.2**	**-1.8**
Knowsley	4.9	5.3	-0.4	-2.8	2.4	2.5	-0.1	2.5	2.7	-0.3
Liverpool	15.3	18.4	-3.2	-7.2	7.2	8.6	-1.4	8.1	9.8	-1.7
St Helens	4.4	4.6	-0.2	-0.9	2.2	2.3	-0.1	2.3	2.3	0.0
Sefton	8.0	7.6	0.4	1.3	3.8	3.8	0.0	4.2	3.8	0.4
Wirral	7.1	7.3	-0.2	-0.6	3.5	3.5	0.0	3.6	3.7	-0.1
YORKSHIRE AND THE HUMBER	**96.1**	**94.9**	**1.2**	**0.2**	**46.0**	**45.5**	**0.5**	**50.2**	**49.5**	**0.7**
East Riding of Yorkshire UA	**17.3**	**12.6**	**4.7**	**15.0**	**8.4**	**6.1**	**2.3**	**9.0**	**6.6**	**2.4**
Kingston upon Hull, City of UA	7.8	10.5	-2.7	-11.3	3.8	5.1	-1.3	3.9	5.4	-1.5
North East Lincolnshire UA	4.2	4.4	-0.2	-1.4	2.1	2.2	-0.1	2.1	2.2	-0.1
North Lincolnshire UA	5.2	4.6	0.6	3.9	2.5	2.2	0.2	2.7	2.4	0.3
York UA	10.2	9.3	0.9	5.1	4.8	4.4	0.4	5.4	4.9	0.5
North Yorkshire	**25.0**	**22.3**	**2.7**	**4.7**	**11.6**	**10.4**	**1.2**	**13.3**	**11.9**	**1.5**
Craven	3.0	2.5	0.5	8.7	1.4	1.3	0.2	1.5	1.2	0.3
Hambleton	4.1	3.8	0.3	3.8	1.9	1.7	0.2	2.2	2.1	0.1
Harrogate	7.1	6.5	0.6	3.8	3.2	2.9	0.3	3.8	3.5	0.3
Richmondshire	3.0	2.8	0.2	3.4	1.2	1.2	0.1	1.7	1.6	0.1
Ryedale	2.6	2.2	0.4	7.0	1.3	1.2	0.1	1.3	1.1	0.3
Scarborough	4.9	4.6	0.3	2.6	2.3	2.3	0.1	2.5	2.3	0.2
Selby	4.0	3.5	0.5	6.6	1.9	1.6	0.3	2.1	1.9	0.2
South Yorkshire (Met County)	**29.1**	**30.8**	**-1.7**	**-1.3**	**14.1**	**15.2**	**-1.0**	**15.0**	**15.6**	**-0.6**
Barnsley	5.5	5.0	0.5	2.4	2.8	2.4	0.3	2.8	2.6	0.2
Doncaster	7.3	7.7	-0.4	-1.3	3.6	3.9	-0.3	3.7	3.8	-0.1
Rotherham	6.5	6.1	0.4	1.6	3.2	3.1	0.1	3.4	3.1	0.3
Sheffield	18.0	20.2	-2.2	-4.3	8.7	9.9	-1.2	9.3	10.3	-1.0
West Yorkshire (Met County)	**47.4**	**50.5**	**-3.1**	**-1.5**	**22.8**	**24.1**	**-1.3**	**24.6**	**26.4**	**-1.8**
Bradford	13.2	15.9	-2.7	-5.8	6.5	7.7	-1.2	6.7	8.2	-1.5
Calderdale	6.3	6.7	-0.4	-2.0	3.1	3.3	-0.2	3.2	3.4	-0.2
Kirklees	13.0	12.6	0.5	1.2	6.3	6.0	0.3	6.7	6.5	0.2
Leeds	27.2	28.4	-1.2	-1.7	12.9	13.4	-0.5	14.3	15.0	-0.8
Wakefield	9.2	8.4	0.8	2.4	4.4	4.1	0.4	4.8	4.3	0.4
EAST MIDLANDS	**112.9**	**96.3**	**16.7**	**4.0**	**54.8**	**46.8**	**7.9**	**58.2**	**49.5**	**8.7**
Derby UA	9.0	9.7	-0.7	-3.1	4.4	4.7	-0.3	4.6	5.0	-0.4
Leicester UA	13.5	15.9	-2.5	-8.8	6.3	7.7	-1.4	7.2	8.2	-1.1
Nottingham UA	18.5	20.2	-1.6	-6.1	9.2	9.8	-0.6	9.4	10.4	-1.0
Rutland UA	2.4	2.2	0.2	6.2	1.2	1.0	0.1	1.3	1.2	0.1
Derbyshire	**24.8**	**21.7**	**3.1**	**4.2**	**12.1**	**10.8**	**1.4**	**12.7**	**11.0**	**1.7**
Amber Valley	4.8	4.3	0.5	4.6	2.4	2.2	0.2	2.5	2.1	0.3
Bolsover	3.8	3.1	0.7	10.4	1.9	1.5	0.4	1.9	1.6	0.4
Chesterfield	3.5	3.3	0.2	1.9	1.7	1.6	0.1	1.8	1.7	0.1
Derbyshire Dales	3.1	3.1	0.0	-0.1	1.5	1.5	0.0	1.6	1.6	0.0
Erewash	4.5	4.1	0.3	3.0	2.2	2.1	0.1	2.3	2.1	0.2
High Peak	3.5	3.5	0.0	0.4	1.7	1.7	0.0	1.8	1.8	0.0
North East Derbyshire	4.3	4.1	0.2	2.4	2.1	2.1	0.0	2.2	2.0	0.2
South Derbyshire	4.8	3.8	1.0	12.5	2.3	1.8	0.5	2.5	2.0	0.5

Note: Bold figures exclude moves between local authorities within each county and moves between counties and unitary authorities within each Government Office Region.
These statistics may be revised in light of the 2001 Census results.

1 Based on patient register data and patient re-registration recorded in the NHSCR.
2 Absolute net migration rate per 1,000 resident population calculated using mid-2001 population estimates based on 2001 Census.

United Kingdom by countries and, within England and Wales Government Office Regions, counties/unitary authorities, districts and London boroughs

thousands

Age 0-15 In	0-15 Out	0-15 Net inflow	16-29 In	16-29 Out	16-29 Net inflow	30-59 In	30-59 Out	30-59 Net inflow	60 and over In	60 and over Out	60 and over Net inflow	Area
7.2	**5.7**	**1.6**	**13.5**	**15.4**	**-1.9**	**13.0**	**10.7**	**2.3**	**3.9**	**2.8**	**1.1**	**Lancashire**
0.8	0.9	-0.2	1.0	1.2	-0.2	1.2	1.4	-0.2	0.3	0.3	0.0	Burnley
0.9	0.8	0.1	1.4	1.4	-0.1	1.8	1.6	0.2	0.4	0.2	0.2	Chorley
1.1	0.8	0.3	1.1	1.4	-0.2	2.0	1.4	0.5	0.8	0.5	0.3	Fylde
0.8	0.8	0.0	1.0	1.1	0.0	1.2	1.2	0.0	0.2	0.2	0.0	Hyndburn
1.1	0.9	0.3	3.8	3.9	-0.1	2.1	1.7	0.4	0.7	0.5	0.2	Lancaster
0.8	0.8	0.0	1.0	1.1	-0.2	1.2	1.2	0.0	0.3	0.3	0.0	Pendle
1.0	1.2	-0.2	3.2	3.3	-0.1	1.8	2.1	-0.4	0.3	0.4	-0.1	Preston
0.7	0.4	0.3	0.7	0.9	-0.2	1.3	0.9	0.4	0.3	0.3	0.1	Ribble Valley
0.6	0.6	0.0	0.8	1.0	-0.1	1.2	1.1	0.1	0.2	0.2	0.0	Rossendale
0.9	0.7	0.2	1.3	1.5	-0.1	1.7	1.6	0.1	0.4	0.3	0.0	South Ribble
1.0	0.6	0.4	1.5	1.7	-0.3	1.7	1.3	0.4	0.4	0.4	0.0	West Lancashire
1.3	0.8	0.5	1.4	1.7	-0.3	2.4	1.7	0.7	1.1	0.7	0.5	Wyre
3.9	**4.1**	**-0.1**	**12.5**	**14.0**	**-1.5**	**7.5**	**8.8**	**-1.3**	**1.6**	**2.3**	**-0.6**	**Merseyside (Met County)**
1.2	1.3	-0.1	1.4	1.6	-0.2	1.9	2.0	0.0	0.4	0.4	0.0	Knowsley
2.1	2.9	-0.8	8.8	8.8	0.0	3.8	5.6	-1.8	0.6	1.2	-0.6	Liverpool
1.0	0.9	0.1	1.5	1.7	-0.2	1.6	1.7	0.0	0.3	0.4	-0.1	St Helens
1.7	1.2	0.5	2.3	2.9	-0.5	3.1	2.7	0.4	0.9	0.8	0.1	Sefton
1.3	1.1	0.2	2.4	3.0	-0.6	2.6	2.5	0.2	0.7	0.7	0.0	Wirral
15.0	**14.1**	**0.9**	**47.5**	**48.5**	**-1.0**	**27.9**	**26.7**	**1.2**	**5.7**	**5.7**	**0.0**	**YORKSHIRE AND THE HUMBER**
3.7	**1.9**	**1.8**	**5.1**	**5.7**	**-0.6**	**6.5**	**3.9**	**2.6**	**2.0**	**1.1**	**0.9**	**East Riding of Yorkshire UA**
1.1	1.9	-0.7	4.1	4.7	-0.6	2.2	3.3	-1.1	0.4	0.7	-0.3	**Kingston upon Hull, City of UA**
1.1	0.9	0.2	1.2	1.7	-0.5	1.5	1.5	0.0	0.4	0.3	0.0	**North East Lincolnshire UA**
1.2	0.9	0.3	1.3	1.8	-0.5	2.1	1.5	0.6	0.5	0.4	0.2	**North Lincolnshire UA**
1.2	1.1	0.0	5.7	4.8	0.9	2.8	2.8	0.0	0.6	0.6	0.0	**York UA**
5.2	**4.2**	**1.0**	**6.9**	**8.6**	**-1.7**	**10.1**	**7.5**	**2.6**	**2.8**	**1.9**	**0.8**	**North Yorkshire**
0.6	0.4	0.2	0.7	0.9	-0.2	1.2	0.9	0.3	0.4	0.3	0.1	Craven
0.8	0.7	0.1	1.0	1.3	-0.3	1.8	1.4	0.3	0.5	0.3	0.2	Hambleton
1.4	1.2	0.2	2.1	2.5	-0.4	2.8	2.1	0.6	0.8	0.6	0.2	Harrogate
0.9	0.9	0.0	0.7	0.8	-0.2	1.1	0.9	0.3	0.2	0.2	0.0	Richmondshire
0.6	0.4	0.2	0.6	0.8	-0.2	1.1	0.8	0.4	0.3	0.3	0.0	Ryedale
0.9	0.8	0.1	1.5	1.8	-0.4	1.8	1.5	0.3	0.7	0.5	0.2	Scarborough
0.9	0.7	0.2	1.0	1.2	-0.1	1.8	1.4	0.3	0.4	0.3	0.1	Selby
4.3	**4.7**	**-0.4**	**15.5**	**15.1**	**0.4**	**7.9**	**9.1**	**-1.2**	**1.4**	**1.9**	**-0.4**	**South Yorkshire (Met County)**
1.2	1.1	0.1	1.8	1.8	0.0	2.2	1.8	0.4	0.4	0.3	0.1	Barnsley
1.5	1.6	0.0	2.4	2.9	-0.5	2.7	2.7	0.0	0.6	0.5	0.0	Doncaster
1.4	1.2	0.3	2.2	2.3	-0.1	2.4	2.3	0.1	0.5	0.4	0.1	Rotherham
1.9	2.7	-0.8	11.6	10.6	1.0	4.0	5.7	-1.7	0.5	1.2	-0.7	Sheffield
6.4	**7.6**	**-1.2**	**26.0**	**24.4**	**1.6**	**12.7**	**15.0**	**-2.3**	**2.3**	**3.4**	**-1.2**	**West Yorkshire (Met County)**
2.4	3.1	-0.7	5.3	6.3	-1.0	4.5	5.3	-0.7	1.0	1.2	-0.2	Bradford
1.3	1.3	0.0	2.0	2.2	-0.2	2.6	2.6	0.0	0.4	0.6	-0.2	Calderdale
2.3	2.2	0.1	5.3	5.4	-0.1	4.5	4.2	0.3	0.9	0.8	0.1	Kirklees
2.8	3.7	-0.9	17.1	14.2	2.9	6.3	8.6	-2.3	0.9	1.8	-0.9	Leeds
1.8	1.6	0.3	3.3	3.3	0.0	3.5	3.0	0.4	0.6	0.5	0.0	Wakefield
19.7	**14.2**	**5.5**	**46.5**	**47.9**	**-1.4**	**37.3**	**28.2**	**9.1**	**9.5**	**6.0**	**3.5**	**EAST MIDLANDS**
1.5	1.7	-0.2	4.3	4.4	-0.1	2.6	3.0	-0.4	0.5	0.5	0.0	**Derby UA**
1.9	2.5	-0.6	7.6	7.9	-0.3	3.3	4.6	-1.3	0.6	0.9	-0.3	**Leicester UA**
1.7	2.8	-1.0	12.5	11.0	1.5	3.7	5.4	-1.7	0.5	1.0	-0.4	**Nottingham UA**
0.6	0.4	0.3	0.6	0.9	-0.3	0.9	0.7	0.2	0.2	0.2	0.1	**Rutland UA**
5.2	**4.1**	**1.0**	**7.2**	**8.0**	**-0.7**	**10.2**	**8.1**	**2.1**	**2.2**	**1.6**	**0.7**	**Derbyshire**
1.0	0.9	0.2	1.4	1.5	-0.1	2.0	1.7	0.4	0.4	0.3	0.1	Amber Valley
0.9	0.8	0.1	1.0	1.0	0.0	1.6	1.1	0.4	0.3	0.2	0.1	Bolsover
0.8	0.6	0.1	1.1	1.1	0.0	1.3	1.3	0.0	0.3	0.3	0.0	Chesterfield
0.6	0.5	0.1	0.8	1.1	-0.3	1.4	1.2	0.2	0.3	0.3	0.1	Derbyshire Dales
0.9	0.8	0.1	1.5	1.4	0.1	1.7	1.6	0.1	0.3	0.3	0.1	Erewash
0.7	0.6	0.0	1.0	1.3	-0.2	1.5	1.4	0.2	0.3	0.2	0.0	High Peak
0.9	0.7	0.1	1.2	1.4	-0.2	1.8	1.6	0.2	0.5	0.3	0.1	North East Derbyshire
1.0	0.7	0.3	1.3	1.4	0.0	2.1	1.4	0.6	0.4	0.2	0.2	South Derbyshire

Table 5.1b - *continued*

Area	Sex									
	Persons				Males			Females		
	In	Out	Net inflow	Rate[2]	In	Out	Net inflow	In	Out	Net inflow
Leicestershire	**25.9**	**22.3**	**3.7**	**6.0**	**12.8**	**11.0**	**1.8**	**13.1**	**11.3**	**1.9**
Blaby	4.8	4.3	0.4	4.9	2.3	2.1	0.2	2.5	2.2	0.3
Charnwood	8.4	8.0	0.4	2.6	4.4	4.1	0.2	4.0	3.9	0.2
Harborough	5.2	3.6	1.6	20.5	2.5	1.7	0.8	2.7	1.9	0.8
Hinckley and Bosworth	4.1	3.9	0.2	2.0	2.0	1.9	0.1	2.1	2.0	0.1
Melton	2.2	2.1	0.1	3.1	1.1	0.9	0.1	1.1	1.1	0.0
North West Leicestershire	4.2	3.4	0.9	10.3	2.1	1.6	0.5	2.1	1.7	0.4
Oadby and Wigston	4.1	4.1	0.0	0.3	2.0	2.0	-0.1	2.2	2.1	0.1
Lincolnshire	**28.7**	**21.1**	**7.6**	**11.8**	**13.7**	**10.0**	**3.7**	**15.0**	**11.1**	**3.9**
Boston	2.7	2.1	0.6	10.8	1.3	1.0	0.3	1.4	1.1	0.3
East Lindsey	8.1	6.3	1.8	14.1	4.0	3.0	1.0	4.1	3.3	0.9
Lincoln	5.7	5.7	-0.1	-0.8	2.7	2.7	0.0	3.0	3.1	-0.1
North Kesteven	6.8	4.5	2.3	24.6	3.1	2.1	1.0	3.7	2.4	1.3
South Holland	4.0	2.7	1.3	17.2	1.9	1.3	0.6	2.1	1.4	0.7
South Kesteven	6.2	5.6	0.6	4.8	3.0	2.7	0.3	3.2	2.9	0.3
West Lindsey	5.3	4.3	1.0	12.8	2.6	2.1	0.5	2.7	2.2	0.5
Northamptonshire	**23.6**	**19.9**	**3.7**	**5.8**	**11.4**	**9.6**	**1.9**	**12.2**	**10.4**	**1.8**
Corby	1.6	1.7	-0.1	-2.3	0.8	0.9	-0.1	0.8	0.8	0.0
Daventry	5.0	3.9	1.1	15.6	2.4	1.8	0.6	2.6	2.0	0.5
East Northamptonshire	5.2	3.5	1.7	21.9	2.6	1.7	0.8	2.6	1.8	0.9
Kettering	4.0	3.1	0.9	10.6	1.9	1.5	0.4	2.0	1.6	0.5
Northampton	8.5	9.5	-1.1	-5.6	4.1	4.5	-0.4	4.4	5.0	-0.6
South Northamptonshire	5.5	4.5	1.0	12.5	2.7	2.2	0.4	2.8	2.3	0.6
Wellingborough	3.5	3.3	0.2	2.9	1.7	1.6	0.1	1.8	1.7	0.1
Nottinghamshire	**28.3**	**25.1**	**3.2**	**4.3**	**13.6**	**12.2**	**1.4**	**14.7**	**12.9**	**1.8**
Ashfield	5.0	4.2	0.8	7.3	2.5	2.1	0.4	2.5	2.2	0.4
Bassetlaw	4.8	3.8	1.0	9.7	2.4	1.9	0.6	2.4	1.9	0.5
Broxtowe	5.9	6.2	-0.2	-2.0	2.8	3.0	-0.1	3.1	3.2	-0.1
Gedling	5.7	5.5	0.2	2.0	2.7	2.6	0.0	3.1	2.8	0.2
Mansfield	3.7	4.3	-0.6	-6.2	1.8	2.1	-0.3	1.9	2.2	-0.3
Newark and Sherwood	5.5	4.7	0.8	7.6	2.6	2.3	0.4	2.9	2.5	0.4
Rushcliffe	7.1	6.0	1.1	10.5	3.3	2.8	0.5	3.8	3.2	0.6
WEST MIDLANDS	**93.5**	**101.0**	**-7.6**	**-1.4**	**44.6**	**49.2**	**-4.6**	**48.9**	**51.9**	**-3.0**
Herefordshire, County of UA	7.0	6.2	0.8	4.3	3.3	2.9	0.4	3.6	3.3	0.4
Stoke-on-Trent UA	8.3	9.7	-1.4	-5.8	4.0	4.7	-0.7	4.3	5.0	-0.7
Telford & Wrekin UA	6.8	5.7	1.0	6.5	3.4	2.9	0.5	3.4	2.9	0.5
Shropshire	**11.3**	**9.9**	**1.3**	**4.7**	**5.4**	**4.7**	**0.7**	**5.8**	**5.2**	**0.7**
Bridgnorth	2.4	2.4	0.0	0.4	1.2	1.2	0.0	1.3	1.2	0.0
North Shropshire	3.3	2.7	0.6	11.0	1.6	1.3	0.3	1.8	1.5	0.3
Oswestry	1.8	1.5	0.3	7.8	0.9	0.7	0.2	0.9	0.8	0.1
Shrewsbury and Atcham	4.1	3.8	0.3	2.7	2.0	1.9	0.1	2.1	1.9	0.2
South Shropshire	2.1	2.0	0.1	3.6	1.1	1.0	0.1	1.1	1.0	0.0
Staffordshire	**28.1**	**25.3**	**2.7**	**3.4**	**13.7**	**12.5**	**1.2**	**14.4**	**12.8**	**1.5**
Cannock Chase	3.3	3.3	0.1	0.8	1.6	1.6	0.0	1.7	1.6	0.1
East Staffordshire	4.7	3.8	1.0	9.2	2.3	1.8	0.4	2.4	1.9	0.5
Lichfield	4.8	4.6	0.2	2.1	2.3	2.3	0.0	2.5	2.3	0.2
Newcastle-under-Lyme	5.7	5.6	0.1	0.5	2.7	2.7	0.0	3.0	2.9	0.1
South Staffordshire	4.9	4.6	0.3	2.9	2.4	2.2	0.3	2.5	2.4	0.0
Stafford	5.5	5.0	0.5	4.1	2.8	2.6	0.2	2.7	2.4	0.3
Staffordshire Moorlands	4.0	3.3	0.7	6.9	1.9	1.6	0.3	2.1	1.7	0.3
Tamworth	3.0	3.0	0.0	-0.2	1.5	1.5	0.0	1.5	1.5	0.0
Warwickshire	**20.6**	**18.4**	**2.2**	**4.3**	**9.4**	**8.9**	**0.5**	**11.1**	**9.5**	**1.6**
North Warwickshire	2.7	3.0	-0.3	-4.7	1.3	1.5	-0.1	1.4	1.5	-0.2
Nuneaton and Bedworth	3.9	3.6	0.2	1.7	1.8	1.8	0.0	2.1	1.9	0.2
Rugby	4.1	3.7	0.4	4.2	1.9	1.8	0.1	2.1	1.9	0.3
Stratford-on-Avon	6.3	5.5	0.8	7.0	2.8	2.6	0.2	3.5	2.9	0.6
Warwick	7.0	5.9	1.1	8.7	3.1	2.8	0.3	3.9	3.1	0.8

Note: Bold figures exclude moves between local authorities within each county and moves between counties and unitary authorities within each Government Office Region.
These statistics may be revised in light of the 2001 Census results.

1 Based on patient register data and patient re-registration recorded in the NHSCR.
2 Absolute net migration rate per 1,000 resident population calculated using mid-2001 population estimates based on 2001 Census.

United Kingdom by countries and, within England and Wales
Government Office Regions, counties/unitary authorities, districts and London boroughs

thousands

Age 0-15 In	0-15 Out	0-15 Net inflow	16-29 In	16-29 Out	16-29 Net inflow	30-59 In	30-59 Out	30-59 Net inflow	60 and over In	60 and over Out	60 and over Net inflow	Area
4.4	**3.1**	**1.3**	**11.1**	**11.1**	**0.0**	**8.6**	**6.6**	**1.9**	**1.9**	**1.5**	**0.5**	**Leicestershire**
1.0	0.8	0.2	1.6	1.5	0.1	1.8	1.7	0.2	0.4	0.4	0.0	Blaby
1.0	1.0	0.0	4.7	4.5	0.3	2.1	2.1	0.0	0.5	0.4	0.1	Charnwood
1.0	0.6	0.4	1.5	1.5	0.0	2.3	1.3	1.0	0.5	0.3	0.2	Harborough
0.8	0.7	0.1	1.3	1.4	-0.1	1.6	1.5	0.1	0.4	0.3	0.1	Hinckley and Bosworth
0.4	0.4	0.1	0.7	0.8	-0.1	1.0	0.8	0.2	0.2	0.2	0.0	Melton
0.9	0.6	0.3	1.1	1.2	0.0	1.8	1.3	0.6	0.3	0.3	0.0	North West Leicestershire
0.7	0.5	0.2	2.1	2.2	-0.1	1.0	1.1	-0.1	0.3	0.3	0.0	Oadby and Wigston
5.8	**3.7**	**2.2**	**7.6**	**8.6**	**-1.0**	**11.0**	**6.7**	**4.3**	**4.3**	**2.2**	**2.1**	**Lincolnshire**
0.6	0.3	0.3	0.7	0.8	-0.1	1.0	0.7	0.3	0.4	0.2	0.1	Boston
1.5	1.2	0.4	1.9	2.2	-0.3	3.1	2.0	1.1	1.6	0.9	0.7	East Lindsey
0.9	1.1	-0.2	2.7	2.4	0.4	1.7	1.9	-0.2	0.4	0.4	0.0	Lincoln
1.6	0.9	0.6	1.4	1.4	-0.1	2.8	1.6	1.1	1.1	0.5	0.6	North Kesteven
0.8	0.4	0.3	0.8	1.0	-0.1	1.6	0.9	0.7	0.8	0.4	0.4	South Holland
1.4	1.0	0.4	1.6	2.1	-0.5	2.5	2.0	0.6	0.7	0.5	0.2	South Kesteven
1.3	0.9	0.4	1.3	1.5	-0.2	2.1	1.4	0.6	0.6	0.4	0.2	West Lindsey
4.9	**3.5**	**1.4**	**7.8**	**8.2**	**-0.3**	**9.1**	**6.8**	**2.3**	**1.8**	**1.5**	**0.3**	**Northamptonshire**
0.4	0.4	0.0	0.5	0.6	-0.1	0.6	0.6	0.0	0.1	0.1	0.0	Corby
1.2	0.8	0.4	1.3	1.2	0.1	2.1	1.5	0.6	0.4	0.4	0.0	Daventry
1.3	0.7	0.6	1.2	1.2	0.0	2.2	1.3	0.9	0.5	0.3	0.2	East Northamptonshire
0.8	0.6	0.2	1.2	1.1	0.0	1.6	1.1	0.5	0.4	0.2	0.1	Kettering
1.3	1.8	-0.4	3.9	4.0	-0.1	2.7	3.2	-0.5	0.5	0.6	0.0	Northampton
1.3	0.8	0.5	1.3	1.5	-0.1	2.5	1.9	0.6	0.4	0.4	0.0	South Northamptonshire
0.9	0.7	0.2	0.9	1.1	-0.2	1.4	1.3	0.2	0.3	0.3	0.0	Wellingborough
5.5	**4.3**	**1.2**	**10.2**	**10.4**	**-0.2**	**10.4**	**8.8**	**1.6**	**2.2**	**1.6**	**0.6**	**Nottinghamshire**
1.1	0.9	0.2	1.5	1.4	0.1	2.0	1.6	0.4	0.4	0.3	0.1	Ashfield
1.1	0.8	0.4	1.3	1.4	0.0	1.9	1.3	0.6	0.4	0.3	0.1	Bassetlaw
1.0	1.1	-0.1	2.5	2.3	0.2	2.1	2.3	-0.3	0.3	0.4	-0.1	Broxtowe
1.1	1.0	0.1	1.9	1.9	0.0	2.2	2.2	0.0	0.5	0.5	0.1	Gedling
0.9	1.0	-0.1	1.2	1.4	-0.3	1.3	1.5	-0.2	0.3	0.3	0.0	Mansfield
1.3	0.9	0.3	1.3	1.8	-0.4	2.3	1.7	0.6	0.6	0.4	0.3	Newark and Sherwood
1.1	0.7	0.4	3.0	2.8	0.2	2.5	2.1	0.4	0.5	0.4	0.1	Rushcliffe
15.3	**15.0**	**0.3**	**43.0**	**49.4**	**-6.3**	**28.8**	**29.8**	**-1.0**	**6.3**	**6.8**	**-0.5**	**WEST MIDLANDS**
1.4	1.0	0.3	1.8	2.4	-0.6	2.9	2.1	0.7	1.0	0.7	0.3	**Herefordshire, County of UA**
1.4	1.8	-0.3	4.0	4.3	-0.3	2.4	3.0	-0.6	0.4	0.6	-0.2	**Stoke-on-Trent UA**
1.6	1.1	0.4	2.3	2.4	-0.1	2.4	1.9	0.5	0.5	0.4	0.1	**Telford & Wrekin UA**
2.4	**1.7**	**0.7**	**3.1**	**4.0**	**-0.9**	**4.5**	**3.3**	**1.1**	**1.3**	**0.9**	**0.4**	**Shropshire**
0.5	0.4	0.0	0.7	0.8	-0.2	1.0	0.9	0.1	0.3	0.2	0.1	Bridgnorth
0.8	0.5	0.3	0.8	1.0	-0.1	1.3	0.9	0.4	0.4	0.3	0.1	North Shropshire
0.4	0.3	0.1	0.5	0.6	-0.1	0.7	0.5	0.2	0.2	0.1	0.1	Oswestry
0.9	0.6	0.3	1.3	1.6	-0.3	1.5	1.3	0.2	0.4	0.3	0.1	Shrewsbury and Atcham
0.4	0.4	0.0	0.4	0.6	-0.2	1.0	0.7	0.2	0.4	0.3	0.1	South Shropshire
5.4	**4.4**	**0.9**	**9.6**	**10.4**	**-0.9**	**10.5**	**8.7**	**1.8**	**2.7**	**1.8**	**0.9**	**Staffordshire**
0.6	0.8	-0.1	1.1	1.0	0.1	1.3	1.2	0.1	0.3	0.3	0.0	Cannock Chase
1.0	0.7	0.3	1.4	1.4	0.0	1.9	1.3	0.6	0.3	0.3	0.1	East Staffordshire
1.0	0.9	0.1	1.3	1.6	-0.3	1.9	1.7	0.1	0.6	0.4	0.2	Lichfield
0.9	0.9	0.0	2.6	2.7	-0.1	1.8	1.7	0.1	0.4	0.4	0.0	Newcastle-under-Lyme
1.0	0.8	0.2	1.2	1.6	-0.3	2.0	1.8	0.2	0.6	0.4	0.2	South Staffordshire
1.0	0.8	0.2	2.1	2.1	-0.1	1.9	1.7	0.2	0.5	0.4	0.1	Stafford
0.9	0.6	0.2	1.0	1.2	-0.2	1.6	1.2	0.4	0.4	0.3	0.2	Staffordshire Moorlands
0.6	0.7	-0.1	1.0	1.0	0.1	1.1	1.2	0.0	0.2	0.2	0.1	Tamworth
3.6	**2.9**	**0.6**	**7.3**	**7.6**	**-0.2**	**8.2**	**6.5**	**1.6**	**1.5**	**1.4**	**0.1**	**Warwickshire**
0.5	0.6	-0.1	0.8	1.0	-0.2	1.2	1.2	0.0	0.2	0.2	0.0	North Warwickshire
0.8	0.7	0.1	1.2	1.4	-0.1	1.5	1.3	0.2	0.3	0.3	0.0	Nuneaton and Bedworth
0.8	0.7	0.1	1.3	1.4	-0.1	1.6	1.4	0.3	0.3	0.3	0.0	Rugby
1.1	0.8	0.3	1.7	2.0	-0.3	2.8	2.2	0.6	0.6	0.5	0.1	Stratford-on-Avon
0.9	0.7	0.1	3.2	2.8	0.4	2.5	2.0	0.6	0.4	0.4	0.0	Warwick

Table 5.1b - *continued*

Area	Sex									
	Persons				Males			Females		
	In	Out	Net inflow	Rate[2]	In	Out	Net inflow	In	Out	Net inflow
West Midlands (Met County)	**51.4**	**66.7**	**-15.4**	**-6.0**	**24.7**	**32.5**	**-7.8**	**26.7**	**34.2**	**-7.6**
Birmingham	30.3	39.3	-9.0	-9.2	14.5	19.0	-4.4	15.8	20.3	-4.5
Coventry	11.3	12.9	-1.7	-5.6	5.5	6.4	-0.9	5.8	6.5	-0.8
Dudley	8.0	8.6	-0.6	-2.0	3.9	4.3	-0.3	4.0	4.3	-0.3
Sandwell	9.1	10.2	-1.2	-4.1	4.4	4.9	-0.5	4.7	5.4	-0.7
Solihull	8.6	8.3	0.3	1.5	4.1	3.9	0.2	4.5	4.3	0.1
Walsall	6.9	8.6	-1.7	-6.7	3.4	4.2	-0.9	3.5	4.3	-0.8
Wolverhampton	7.4	8.9	-1.5	-6.4	3.5	4.4	-0.9	3.8	4.5	-0.7
Worcestershire	**18.5**	**17.3**	**1.2**	**2.2**	**9.1**	**8.5**	**0.6**	**9.4**	**8.9**	**0.6**
Bromsgrove	4.4	3.8	0.6	6.6	2.1	1.8	0.2	2.3	2.0	0.4
Malvern Hills	4.2	3.8	0.4	4.9	2.0	1.9	0.1	2.2	2.0	0.2
Redditch	2.7	3.1	-0.4	-5.1	1.3	1.5	-0.2	1.4	1.6	-0.2
Worcester	4.5	4.8	-0.3	-3.1	2.2	2.3	0.0	2.3	2.5	-0.2
Wychavon	5.9	4.9	1.0	8.8	2.9	2.4	0.5	3.0	2.5	0.5
Wyre Forest	3.3	3.4	-0.1	-0.6	1.7	1.7	0.0	1.6	1.7	-0.1
EAST	**144.9**	**125.4**	**19.5**	**3.6**	**70.3**	**60.9**	**9.4**	**74.6**	**64.5**	**10.1**
Luton UA	6.5	9.1	-2.6	-13.8	3.3	4.5	-1.2	3.3	4.6	-1.4
Peterborough UA	7.0	6.9	0.1	0.6	3.4	3.5	0.0	3.6	3.4	0.1
Southend-on-Sea UA	7.6	6.9	0.7	4.3	3.7	3.4	0.4	3.9	3.5	0.3
Thurrock UA	5.5	5.5	0.0	0.0	2.6	2.7	-0.1	2.9	2.8	0.1
Bedfordshire	**17.8**	**16.6**	**1.2**	**3.1**	**8.7**	**8.2**	**0.5**	**9.1**	**8.4**	**0.7**
Bedford	6.5	6.1	0.4	2.9	3.2	3.0	0.2	3.3	3.1	0.2
Mid Bedfordshire	7.4	6.8	0.6	4.8	3.6	3.4	0.3	3.8	3.5	0.3
South Bedfordshire	6.0	5.8	0.2	1.5	2.9	2.9	0.0	3.1	2.9	0.2
Cambridgeshire	**28.7**	**25.5**	**3.2**	**5.8**	**14.1**	**12.4**	**1.7**	**14.6**	**13.1**	**1.5**
Cambridge	10.8	11.6	-0.8	-7.2	5.4	5.7	-0.4	5.5	5.9	-0.4
East Cambridgeshire	4.3	3.4	0.9	12.3	2.1	1.7	0.4	2.2	1.8	0.5
Fenland	4.8	3.7	1.0	12.3	2.3	1.8	0.5	2.4	1.9	0.5
Huntingdonshire	7.9	7.3	0.6	3.8	3.9	3.5	0.4	4.0	3.8	0.2
South Cambridgeshire	8.9	7.5	1.4	11.0	4.3	3.6	0.7	4.6	3.9	0.7
Essex	**45.6**	**39.1**	**6.5**	**4.9**	**22.1**	**18.9**	**3.2**	**23.5**	**20.2**	**3.3**
Basildon	7.1	6.9	0.2	1.4	3.5	3.4	0.1	3.6	3.4	0.1
Braintree	7.0	5.5	1.5	11.6	3.4	2.7	0.7	3.6	2.7	0.8
Brentwood	3.6	3.6	0.0	0.1	1.8	1.8	0.0	1.8	1.8	0.0
Castle Point	3.7	3.4	0.2	2.8	1.8	1.6	0.1	1.9	1.8	0.1
Chelmsford	7.7	7.3	0.4	2.8	3.7	3.5	0.2	4.1	3.8	0.3
Colchester	8.4	7.8	0.5	3.4	4.0	3.6	0.4	4.4	4.2	0.2
Epping Forest	6.3	6.3	0.0	0.0	3.1	3.1	0.0	3.3	3.3	0.0
Harlow	3.2	3.4	-0.3	-3.2	1.5	1.7	-0.1	1.6	1.8	-0.1
Maldon	3.6	2.7	0.9	15.2	1.7	1.3	0.4	1.9	1.3	0.5
Rochford	3.8	3.3	0.6	7.3	1.8	1.5	0.3	2.0	1.7	0.3
Tendring	6.9	4.8	2.2	15.7	3.4	2.3	1.1	3.5	2.5	1.0
Uttlesford	4.2	4.1	0.1	0.8	2.0	1.9	0.0	2.2	2.2	0.0
Hertfordshire	**41.2**	**41.4**	**-0.2**	**-0.1**	**19.8**	**19.9**	**-0.1**	**21.4**	**21.4**	**0.0**
Broxbourne	4.6	4.2	0.4	4.9	2.2	2.1	0.1	2.4	2.1	0.3
Dacorum	5.9	6.4	-0.5	-3.4	2.8	3.1	-0.3	3.1	3.3	-0.2
East Hertfordshire	7.1	6.7	0.4	3.0	3.4	3.3	0.1	3.7	3.4	0.3
Hertsmere	5.1	5.4	-0.3	-3.7	2.4	2.6	-0.2	2.7	2.9	-0.2
North Hertfordshire	6.3	5.7	0.6	4.9	3.0	2.8	0.2	3.2	2.9	0.3
St Albans	6.9	7.0	-0.1	-0.9	3.4	3.4	0.1	3.4	3.6	-0.2
Stevenage	3.7	3.6	0.1	0.9	1.8	1.8	0.1	1.9	1.8	0.0
Three Rivers	4.7	4.7	0.0	0.3	2.2	2.2	0.0	2.5	2.5	0.0
Watford	4.6	5.0	-0.4	-5.1	2.3	2.4	-0.1	2.4	2.6	-0.3
Welwyn Hatfield	5.7	6.0	-0.3	-3.0	2.7	2.8	-0.1	3.0	3.1	-0.2

Note: Bold figures exclude moves between local authorities within each county and moves between counties and unitary authorities within each Government Office Region.
These statistics may be revised in light of the 2001 Census results.

1 Based on patient register data and patient re-registration recorded in the NHSCR.
2 Absolute net migration rate per 1,000 resident population calculated using mid-2001 population estimates based on 2001 Census.

United Kingdom by countries and, within England and Wales Government Office Regions, counties/unitary authorities, districts and London boroughs

thousands

Age 0-15 In	0-15 Out	0-15 Net inflow	16-29 In	16-29 Out	16-29 Net inflow	30-59 In	30-59 Out	30-59 Net inflow	60 and over In	60 and over Out	60 and over Net inflow	Area
7.6	**10.5**	**-3.0**	**27.6**	**29.9**	**-2.3**	**13.9**	**21.1**	**-7.2**	**2.3**	**5.2**	**-2.8**	**West Midlands (Met County)**
4.3	6.7	-2.4	16.4	16.8	-0.4	8.2	13.0	-4.7	1.3	2.8	-1.5	Birmingham
1.4	1.9	-0.5	7.0	6.9	0.1	2.5	3.5	-1.0	0.4	0.7	-0.4	Coventry
1.7	1.6	0.1	2.6	3.0	-0.4	3.0	3.2	-0.2	0.6	0.7	-0.1	Dudley
2.1	2.3	-0.2	3.3	3.4	-0.1	3.2	3.8	-0.6	0.5	0.7	-0.3	Sandwell
1.8	1.3	0.5	2.6	3.0	-0.5	3.4	3.1	0.4	0.7	0.9	-0.2	Solihull
1.6	1.8	-0.2	2.3	2.9	-0.6	2.5	3.1	-0.6	0.4	0.8	-0.3	Walsall
1.4	1.8	-0.3	3.0	3.6	-0.6	2.5	3.0	-0.5	0.4	0.5	-0.1	Wolverhampton
3.5	**2.9**	**0.6**	**5.7**	**6.8**	**-1.1**	**7.3**	**6.3**	**1.0**	**2.1**	**1.4**	**0.7**	**Worcestershire**
0.9	0.6	0.3	1.1	1.3	-0.3	1.9	1.5	0.4	0.5	0.4	0.1	Bromsgrove
0.9	0.6	0.3	1.0	1.5	-0.5	1.6	1.3	0.3	0.7	0.4	0.3	Malvern Hills
0.5	0.6	-0.1	0.9	1.0	-0.1	1.0	1.2	-0.1	0.2	0.2	0.0	Redditch
0.7	0.9	-0.3	1.9	1.7	0.2	1.6	1.8	-0.2	0.3	0.3	0.0	Worcester
1.1	0.8	0.3	1.5	1.6	-0.1	2.6	1.9	0.7	0.7	0.5	0.2	Wychavon
0.6	0.6	0.0	1.0	1.2	-0.2	1.3	1.3	0.0	0.4	0.3	0.1	Wyre Forest
24.5	**19.5**	**5.0**	**54.5**	**54.7**	**-0.3**	**52.2**	**42.0**	**10.3**	**13.7**	**9.2**	**4.5**	**EAST**
1.2	1.8	-0.7	2.8	3.4	-0.7	2.3	3.3	-1.0	0.4	0.5	-0.2	**Luton UA**
1.3	1.4	0.0	2.7	2.4	0.3	2.4	2.7	-0.2	0.5	0.5	0.0	**Peterborough UA**
1.4	1.3	0.0	2.4	2.3	0.2	2.8	2.6	0.2	1.0	0.7	0.3	**Southend-on-Sea UA**
1.1	1.3	-0.2	1.9	1.7	0.2	2.2	2.2	0.0	0.3	0.4	0.0	**Thurrock UA**
3.4	**3.0**	**0.4**	**6.0**	**6.1**	**-0.1**	**7.2**	**6.3**	**0.9**	**1.2**	**1.3**	**0.0**	**Bedfordshire**
1.2	1.0	0.2	2.5	2.6	-0.1	2.3	2.0	0.3	0.5	0.4	0.1	Bedford
1.4	1.3	0.1	2.2	2.1	0.1	3.3	2.8	0.5	0.5	0.5	-0.1	Mid Bedfordshire
1.2	1.1	0.1	1.9	1.9	0.0	2.5	2.4	0.1	0.4	0.5	0.0	South Bedfordshire
4.4	**3.7**	**0.7**	**12.6**	**11.9**	**0.7**	**9.6**	**8.3**	**1.3**	**2.1**	**1.7**	**0.5**	**Cambridgeshire**
0.7	1.0	-0.3	7.6	7.0	0.6	2.2	3.1	-0.9	0.3	0.5	-0.2	Cambridge
0.8	0.6	0.2	1.2	1.1	0.1	1.8	1.4	0.5	0.4	0.3	0.1	East Cambridgeshire
1.0	0.8	0.2	1.1	1.1	0.0	1.9	1.4	0.5	0.7	0.4	0.3	Fenland
1.6	1.5	0.1	2.3	2.4	-0.1	3.3	2.9	0.4	0.7	0.6	0.1	Huntingdonshire
1.7	1.3	0.4	3.0	2.7	0.2	3.6	2.9	0.8	0.6	0.5	0.1	South Cambridgeshire
8.7	**6.6**	**2.0**	**14.5**	**14.8**	**-0.3**	**17.5**	**14.0**	**3.6**	**4.9**	**3.7**	**1.2**	**Essex**
1.4	1.4	0.0	2.3	2.2	0.1	2.7	2.7	0.0	0.7	0.6	0.1	Basildon
1.4	1.0	0.4	1.8	1.7	0.1	3.0	2.2	0.9	0.7	0.5	0.2	Braintree
0.7	0.5	0.2	1.2	1.3	-0.1	1.4	1.5	-0.1	0.3	0.4	-0.1	Brentwood
0.8	0.5	0.2	0.8	1.2	-0.3	1.5	1.3	0.2	0.6	0.5	0.2	Castle Point
1.3	1.1	0.3	2.9	2.8	0.1	2.9	2.7	0.1	0.6	0.7	-0.1	Chelmsford
1.6	1.7	-0.1	3.4	3.2	0.3	2.7	2.5	0.2	0.7	0.5	0.2	Colchester
1.1	1.1	0.0	1.9	1.9	0.0	2.8	2.6	0.1	0.6	0.7	-0.1	Epping Forest
0.6	0.7	-0.1	1.2	1.0	0.2	1.2	1.3	-0.1	0.2	0.3	-0.2	Harlow
0.7	0.5	0.2	0.8	0.9	0.0	1.6	1.1	0.5	0.4	0.3	0.2	Maldon
0.8	0.5	0.3	1.0	1.1	-0.1	1.6	1.2	0.3	0.5	0.5	0.1	Rochford
1.4	0.9	0.5	1.3	1.5	-0.3	2.8	1.6	1.2	1.5	0.7	0.7	Tendring
0.9	0.7	0.1	1.2	1.4	-0.2	1.7	1.6	0.1	0.4	0.4	0.0	Uttlesford
6.5	**6.3**	**0.2**	**16.1**	**15.5**	**0.6**	**15.5**	**16.1**	**-0.6**	**3.1**	**3.5**	**-0.4**	**Hertfordshire**
1.0	0.8	0.2	1.4	1.2	0.2	1.9	1.8	0.1	0.4	0.4	-0.1	Broxbourne
1.0	1.1	0.0	2.0	2.1	-0.1	2.4	2.6	-0.2	0.5	0.5	0.0	Dacorum
1.3	1.1	0.2	2.5	2.4	0.1	2.8	2.8	0.1	0.5	0.5	0.0	East Hertfordshire
0.9	1.0	-0.1	1.7	1.8	-0.1	2.0	2.2	-0.2	0.5	0.5	0.0	Hertsmere
1.0	0.9	0.1	2.1	2.1	0.0	2.6	2.3	0.3	0.6	0.4	0.2	North Hertfordshire
0.9	0.9	0.0	2.9	2.7	0.2	2.7	2.7	-0.1	0.4	0.6	-0.3	St Albans
0.7	0.7	-0.1	1.4	1.2	0.2	1.4	1.4	0.0	0.2	0.3	-0.1	Stevenage
0.9	0.8	0.1	1.4	1.5	-0.1	1.9	1.9	0.1	0.5	0.5	0.0	Three Rivers
0.6	0.8	-0.2	2.0	1.8	0.2	1.7	2.1	-0.4	0.3	0.3	0.0	Watford
0.8	0.8	0.0	2.6	2.6	0.0	1.9	2.1	-0.2	0.4	0.5	-0.1	Welwyn Hatfield

Table 5.1b - *continued*

Area	Persons In	Persons Out	Persons Net inflow	Persons Rate[2]	Males In	Males Out	Males Net inflow	Females In	Females Out	Females Net inflow
Norfolk	**27.7**	**21.1**	**6.5**	**8.2**	**13.4**	**10.2**	**3.2**	**14.2**	**10.9**	**3.3**
Breckland	6.4	5.2	1.2	9.7	3.1	2.5	0.6	3.3	2.7	0.6
Broadland	6.6	5.2	1.4	12.2	3.1	2.5	0.6	3.5	2.7	0.8
Great Yarmouth	4.1	3.5	0.7	7.4	2.1	1.8	0.3	2.1	1.7	0.4
King's Lynn and West Norfolk	6.1	5.0	1.1	8.0	3.0	2.4	0.6	3.1	2.6	0.5
North Norfolk	5.0	3.7	1.3	12.8	2.4	1.8	0.6	2.6	2.0	0.6
Norwich	8.8	8.7	0.1	1.2	4.3	4.1	0.2	4.5	4.5	0.0
South Norfolk	6.6	5.8	0.8	6.8	3.2	2.8	0.3	3.4	3.0	0.4
Suffolk	**22.2**	**18.2**	**4.1**	**6.1**	**10.8**	**8.9**	**1.9**	**11.4**	**9.2**	**2.2**
Babergh	4.2	3.8	0.4	5.0	2.0	1.9	0.1	2.2	1.9	0.3
Forest Heath	2.6	2.3	0.3	5.0	1.3	1.1	0.1	1.4	1.2	0.1
Ipswich	4.9	4.9	0.0	0.0	2.4	2.4	0.0	2.5	2.5	0.0
Mid Suffolk	4.8	3.9	0.9	10.1	2.3	1.8	0.5	2.5	2.1	0.4
St Edmundsbury	4.8	4.2	0.6	5.8	2.3	2.1	0.2	2.5	2.1	0.3
Suffolk Coastal	5.5	4.8	0.7	6.3	2.6	2.3	0.3	2.9	2.5	0.4
Waveney	4.8	3.6	1.2	10.4	2.4	1.8	0.6	2.4	1.8	0.6
LONDON	**163.6**	**232.2**	**-68.6**	**-9.5**	**79.7**	**111.1**	**-31.4**	**83.9**	**121.1**	**-37.2**
Inner London	**121.4**	**160.4**	**-39.0**	**-14.1**	**57.9**	**75.1**	**-17.2**	**63.6**	**85.3**	**-21.8**
Camden	17.2	19.0	-1.8	-9.1	7.8	8.3	-0.5	9.4	10.7	-1.3
City of London	0.8	0.7	0.1	13.7	0.4	0.3	0.1	0.4	0.4	0.0
Hackney	13.5	16.7	-3.2	-16.0	6.3	8.0	-1.8	7.2	8.7	-1.5
Hammersmith and Fulham	13.4	15.9	-2.5	-15.1	5.9	6.7	-0.8	7.5	9.2	-1.7
Haringey	16.0	22.0	-6.0	-27.7	7.3	10.3	-3.0	8.7	11.7	-3.0
Islington	15.0	17.0	-2.0	-11.6	6.9	7.7	-0.8	8.0	9.3	-1.3
Kensington and Chelsea	9.8	12.1	-2.3	-14.2	4.4	5.4	-1.1	5.4	6.6	-1.2
Lambeth	22.4	25.7	-3.3	-12.3	10.5	11.8	-1.3	11.9	13.9	-2.0
Lewisham	16.0	18.4	-2.4	-9.8	7.6	8.6	-1.1	8.4	9.8	-1.4
Newham	12.6	18.0	-5.4	-22.1	6.1	8.9	-2.8	6.6	9.1	-2.6
Southwark	17.9	20.8	-2.8	-11.6	8.6	9.6	-1.0	9.3	11.2	-1.8
Tower Hamlets	12.4	14.1	-1.6	-8.2	6.2	6.7	-0.6	6.3	7.3	-1.1
Wandsworth	23.8	27.2	-3.4	-13.1	10.3	12.1	-1.8	13.5	15.1	-1.6
Westminster	16.1	18.3	-2.2	-12.3	7.7	8.5	-0.9	8.4	9.8	-1.4
Outer London	**164.1**	**193.7**	**-29.6**	**-6.7**	**78.3**	**92.5**	**-14.1**	**85.8**	**101.2**	**-15.5**
Barking and Dagenham	8.0	8.3	-0.3	-1.6	3.8	4.0	-0.2	4.2	4.3	-0.1
Barnet	19.2	20.7	-1.5	-4.6	9.1	9.5	-0.4	10.1	11.1	-1.0
Bexley	9.3	9.9	-0.6	-2.8	4.3	4.8	-0.4	4.9	5.1	-0.2
Brent	14.9	20.9	-6.0	-22.8	7.0	10.0	-3.0	7.8	10.9	-3.1
Bromley	13.7	14.2	-0.5	-1.8	6.6	6.8	-0.3	7.1	7.4	-0.2
Croydon	15.7	18.1	-2.4	-7.1	7.6	8.8	-1.2	8.1	9.3	-1.2
Ealing	17.7	23.1	-5.4	-18.0	8.5	11.1	-2.6	9.2	12.0	-2.8
Enfield	15.0	16.5	-1.5	-5.4	7.2	8.0	-0.8	7.8	8.5	-0.7
Greenwich	13.8	14.1	-0.3	-1.2	6.6	6.7	-0.1	7.2	7.4	-0.2
Harrow	12.1	13.3	-1.3	-6.1	5.9	6.5	-0.6	6.1	6.9	-0.7
Havering	8.3	8.1	0.2	1.0	3.9	3.9	0.0	4.4	4.2	0.2
Hillingdon	13.2	14.3	-1.0	-4.3	6.4	6.8	-0.4	6.8	7.4	-0.6
Hounslow	13.1	16.1	-3.0	-14.1	6.4	7.6	-1.2	6.7	8.5	-1.8
Kingston upon Thames	10.6	11.0	-0.3	-2.2	5.1	5.2	-0.1	5.6	5.8	-0.2
Merton	12.1	13.8	-1.7	-9.1	5.6	6.5	-0.9	6.5	7.3	-0.8
Redbridge	13.4	13.9	-0.5	-2.1	6.4	6.5	-0.1	6.9	7.3	-0.4
Richmond upon Thames	12.3	12.8	-0.5	-2.8	5.7	5.9	-0.2	6.7	6.9	-0.3
Sutton	9.4	9.3	0.1	0.6	4.5	4.5	0.0	5.0	4.8	0.2
Waltham Forest	10.9	14.1	-3.2	-14.7	5.3	6.9	-1.7	5.6	7.2	-1.5

Note: Bold figures exclude moves between local authorities within each county and moves between counties and unitary authorities within each Government Office Region.
These statistics may be revised in light of the 2001 Census results.

1 Based on patient register data and patient re-registration recorded in the NHSCR.
2 Absolute net migration rate per 1,000 resident population calculated using mid-2001 population estimates based on 2001 Census.

**United Kingdom by countries and, within England and Wales
Government Office Regions, counties/unitary authorities, districts and London boroughs**

thousands

Age 0-15			16-29			30-59			60 and over			Area
In	Out	Net inflow	In	Out	Net inflow	In	Out	Net inflow	In	Out	Net inflow	
4.9	**3.5**	**1.4**	**8.4**	**8.8**	**-0.4**	**10.4**	**6.8**	**3.6**	**3.9**	**2.0**	**1.9**	**Norfolk**
1.4	1.1	0.3	1.5	1.7	-0.3	2.5	1.8	0.7	1.0	0.5	0.5	Breckland
1.2	0.9	0.3	1.8	1.8	0.0	2.7	2.0	0.7	0.9	0.5	0.4	Broadland
0.9	0.7	0.1	0.9	1.1	-0.2	1.7	1.3	0.4	0.6	0.3	0.3	Great Yarmouth
1.1	0.9	0.3	1.5	1.8	-0.3	2.5	1.7	0.8	1.0	0.6	0.4	King's Lynn and West Norfolk
0.9	0.6	0.3	1.0	1.2	-0.2	2.1	1.3	0.8	1.0	0.6	0.4	North Norfolk
0.9	1.2	-0.3	5.0	4.0	1.0	2.5	3.0	-0.5	0.4	0.5	-0.1	Norwich
1.4	1.0	0.4	1.6	2.0	-0.4	2.8	2.2	0.7	0.7	0.6	0.1	South Norfolk
4.3	**3.2**	**1.1**	**6.3**	**7.2**	**-0.9**	**8.8**	**6.2**	**2.6**	**2.9**	**1.6**	**1.3**	**Suffolk**
0.9	0.7	0.1	1.1	1.4	-0.3	1.7	1.4	0.4	0.6	0.4	0.2	Babergh
0.4	0.4	0.0	0.9	0.8	0.1	1.0	0.9	0.1	0.3	0.2	0.1	Forest Heath
0.9	0.8	0.1	2.1	1.9	0.2	1.7	1.7	-0.1	0.3	0.4	-0.1	Ipswich
1.0	0.7	0.3	1.2	1.3	-0.2	2.0	1.4	0.6	0.6	0.4	0.1	Mid Suffolk
1.0	0.8	0.1	1.4	1.4	-0.1	1.9	1.6	0.3	0.5	0.4	0.2	St Edmundsbury
1.1	0.9	0.3	1.3	1.8	-0.5	2.3	1.7	0.6	0.9	0.5	0.4	Suffolk Coastal
0.9	0.7	0.2	1.1	1.3	-0.1	1.9	1.2	0.7	0.8	0.4	0.4	Waveney
12.6	**41.6**	**-29.0**	**101.4**	**74.8**	**26.7**	**44.0**	**94.5**	**-50.5**	**5.5**	**21.3**	**-15.7**	**LONDON**
9.7	**28.1**	**-18.4**	**72.8**	**53.7**	**19.1**	**35.9**	**67.4**	**-31.5**	**3.0**	**11.3**	**-8.2**	**Inner London**
1.1	2.0	-0.9	10.2	8.5	1.7	5.4	7.5	-2.1	0.5	1.0	-0.5	Camden
0.0	0.1	0.0	0.4	0.3	0.1	0.4	0.3	0.0	0.0	0.1	0.0	City of London
1.5	3.1	-1.6	6.4	5.6	0.8	5.2	7.1	-1.9	0.3	1.0	-0.7	Hackney
1.0	1.9	-0.9	7.6	6.2	1.4	4.6	7.0	-2.5	0.3	0.8	-0.5	Hammersmith and Fulham
2.1	4.0	-1.9	7.6	7.6	0.0	5.8	9.3	-3.5	0.5	1.2	-0.7	Haringey
1.2	2.3	-1.1	8.3	6.5	1.9	5.1	7.4	-2.3	0.3	0.8	-0.5	Islington
0.7	1.5	-0.8	5.0	4.3	0.7	3.8	5.5	-1.8	0.4	0.8	-0.4	Kensington and Chelsea
1.7	4.0	-2.3	12.5	9.5	3.1	7.7	11.0	-3.3	0.5	1.3	-0.8	Lambeth
2.0	3.3	-1.3	7.4	6.2	1.2	6.0	7.8	-1.8	0.6	1.2	-0.6	Lewisham
2.5	4.4	-1.9	5.2	5.7	-0.5	4.5	6.8	-2.4	0.5	1.2	-0.7	Newham
1.5	3.3	-1.8	9.5	7.7	1.8	6.5	8.6	-2.1	0.4	1.2	-0.7	Southwark
1.0	2.2	-1.3	7.0	5.4	1.6	4.2	5.5	-1.3	0.3	0.9	-0.7	Tower Hamlets
1.6	3.5	-1.9	14.6	10.4	4.3	7.0	11.9	-4.9	0.5	1.4	-0.9	Wandsworth
1.4	2.2	-0.8	8.3	7.3	1.0	5.8	7.7	-1.9	0.5	1.1	-0.6	Westminster
22.7	**33.3**	**-10.6**	**74.5**	**67.0**	**7.6**	**58.1**	**77.2**	**-19.1**	**8.7**	**16.3**	**-7.5**	**Outer London**
1.8	2.0	-0.1	2.7	2.4	0.3	3.1	3.2	-0.1	0.4	0.7	-0.3	Barking and Dagenham
2.8	3.1	-0.4	8.7	8.1	0.6	6.7	8.0	-1.3	1.0	1.5	-0.4	Barnet
1.8	1.9	0.0	3.0	3.1	-0.1	3.6	3.9	-0.3	0.8	1.0	-0.2	Bexley
1.8	3.4	-1.6	6.9	7.7	-0.8	5.5	8.5	-3.0	0.6	1.3	-0.7	Brent
2.2	2.3	-0.1	5.0	4.7	0.4	5.4	5.8	-0.5	1.0	1.4	-0.4	Bromley
2.5	3.3	-0.7	6.1	5.9	0.2	6.1	7.5	-1.4	1.0	1.4	-0.4	Croydon
2.0	3.7	-1.7	8.8	8.4	0.4	6.3	9.6	-3.3	0.6	1.4	-0.8	Ealing
2.7	3.2	-0.5	5.7	5.4	0.3	5.6	6.5	-0.9	1.0	1.4	-0.4	Enfield
2.1	2.6	-0.5	6.0	5.0	1.0	5.0	5.4	-0.4	0.6	1.0	-0.3	Greenwich
2.2	2.2	0.0	4.6	4.5	0.0	4.4	5.3	-0.8	0.8	1.3	-0.5	Harrow
1.7	1.3	0.4	2.5	2.6	-0.1	3.3	3.2	0.1	0.8	0.9	-0.1	Havering
2.1	2.6	-0.6	5.9	5.1	0.8	4.5	5.4	-0.9	0.7	1.1	-0.4	Hillingdon
1.7	2.9	-1.2	5.9	5.5	0.4	4.9	6.6	-1.6	0.5	1.1	-0.5	Hounslow
1.1	1.5	-0.4	5.5	4.5	1.0	3.5	4.2	-0.7	0.5	0.7	-0.2	Kingston upon Thames
1.5	2.5	-1.0	5.5	4.6	0.9	4.6	5.9	-1.3	0.6	0.9	-0.4	Merton
2.7	2.6	0.1	4.8	4.7	0.2	4.9	5.3	-0.4	0.9	1.3	-0.3	Redbridge
1.2	1.7	-0.5	5.7	4.6	1.1	4.9	5.5	-0.6	0.5	1.0	-0.5	Richmond upon Thames
1.5	1.7	-0.2	3.5	2.9	0.7	3.8	3.9	-0.2	0.7	0.8	-0.2	Sutton
1.5	3.0	-1.5	4.5	4.2	0.3	4.4	5.9	-1.5	0.5	1.0	-0.5	Waltham Forest

Table 5.1b - *continued*

Area	Persons In	Persons Out	Persons Net inflow	Persons Rate[2]	Males In	Males Out	Males Net inflow	Females In	Females Out	Females Net inflow
SOUTH EAST	**220.3**	**212.9**	**7.5**	**0.9**	**106.8**	**103.1**	**3.7**	**113.5**	**109.7**	**3.8**
Bracknell Forest UA	6.0	6.9	-0.9	-7.9	3.0	3.4	-0.4	3.0	3.5	-0.5
Brighton and Hove UA	16.3	15.5	0.8	3.2	8.0	7.4	0.5	8.3	8.0	0.3
Isle of Wight UA	5.5	4.0	1.5	11.5	2.7	1.9	0.8	2.8	2.1	0.8
Medway UA	9.9	9.9	0.0	0.2	4.8	4.9	-0.1	5.1	5.0	0.1
Milton Keynes UA	10.9	9.4	1.5	7.3	5.5	4.7	0.8	5.4	4.7	0.7
Portsmouth UA	9.6	9.9	-0.3	-1.5	4.8	5.0	-0.3	4.8	4.8	0.0
Reading UA	10.0	10.6	-0.6	-4.4	5.0	5.2	-0.2	5.0	5.4	-0.4
Slough UA	5.4	6.7	-1.3	-11.1	2.6	3.2	-0.6	2.8	3.5	-0.7
Southampton UA	13.2	13.5	-0.3	-1.4	6.7	6.9	-0.3	6.5	6.5	0.0
West Berkshire UA	7.9	8.4	-0.6	-4.1	3.8	4.1	-0.3	4.1	4.3	-0.3
Windsor and Maidenhead UA	7.8	8.9	-1.1	-8.1	3.8	4.4	-0.5	4.0	4.5	-0.5
Wokingham UA	9.4	9.8	-0.5	-3.0	4.5	4.7	-0.2	4.8	5.1	-0.2
Buckinghamshire	**21.0**	**22.7**	**-1.7**	**-3.5**	**9.9**	**10.9**	**-1.0**	**11.1**	**11.8**	**-0.7**
Aylesbury Vale	8.4	8.2	0.1	0.8	4.0	4.0	0.0	4.3	4.2	0.1
Chiltern	4.8	5.1	-0.3	-3.1	2.2	2.4	-0.2	2.6	2.7	-0.1
South Bucks	4.0	4.2	-0.1	-2.1	1.9	2.0	-0.1	2.1	2.2	-0.1
Wycombe	7.8	9.2	-1.4	-8.5	3.6	4.4	-0.7	4.2	4.8	-0.6
East Sussex	**21.3**	**16.6**	**4.7**	**9.6**	**10.3**	**8.0**	**2.3**	**11.0**	**8.5**	**2.4**
Eastbourne	6.0	4.5	1.5	16.2	2.8	2.2	0.6	3.2	2.3	0.8
Hastings	4.6	3.9	0.7	8.1	2.3	2.0	0.3	2.3	1.9	0.4
Lewes	5.1	4.4	0.7	7.4	2.4	2.1	0.3	2.7	2.3	0.4
Rother	5.4	4.5	0.8	9.7	2.7	2.2	0.5	2.7	2.4	0.4
Wealden	8.2	7.1	1.1	7.7	4.0	3.4	0.6	4.2	3.7	0.5
Hampshire	**50.4**	**49.5**	**0.8**	**0.7**	**24.2**	**23.7**	**0.5**	**26.1**	**25.8**	**0.3**
Basingstoke and Deane	6.8	7.4	-0.7	-4.4	3.4	3.7	-0.4	3.4	3.7	-0.3
East Hampshire	6.8	6.5	0.3	2.8	3.2	3.1	0.1	3.6	3.4	0.2
Eastleigh	6.1	5.9	0.3	2.3	3.1	2.8	0.2	3.1	3.0	0.1
Fareham	5.9	5.2	0.7	6.5	2.8	2.5	0.3	3.0	2.7	0.4
Gosport	3.5	3.7	-0.3	-3.8	1.6	1.7	-0.1	1.9	2.0	-0.1
Hart	5.6	5.5	0.0	0.5	2.6	2.6	0.0	2.9	2.9	0.0
Havant	5.6	5.7	-0.1	-1.1	2.7	2.8	0.0	2.8	2.9	-0.1
New Forest	8.3	7.2	1.1	6.4	4.0	3.5	0.5	4.3	3.7	0.6
Rushmoor	5.4	6.6	-1.2	-13.5	2.6	3.1	-0.5	2.8	3.6	-0.7
Test Valley	6.3	5.7	0.6	5.1	3.0	2.8	0.3	3.2	2.9	0.3
Winchester	7.7	7.5	0.2	2.0	3.6	3.6	0.1	4.1	4.0	0.2
Kent	**46.0**	**41.3**	**4.8**	**3.6**	**22.5**	**20.2**	**2.3**	**23.5**	**21.1**	**2.4**
Ashford	6.5	5.0	1.5	14.8	3.2	2.4	0.8	3.3	2.6	0.8
Canterbury	9.1	8.5	0.6	4.1	4.3	4.0	0.3	4.8	4.5	0.2
Dartford	4.6	4.8	-0.2	-2.3	2.2	2.3	-0.1	2.3	2.4	-0.1
Dover	4.7	4.3	0.4	3.5	2.3	2.2	0.1	2.4	2.1	0.3
Gravesham	3.6	3.8	-0.2	-1.8	1.8	1.9	-0.1	1.8	1.9	-0.1
Maidstone	7.3	6.9	0.4	3.0	3.6	3.4	0.3	3.7	3.5	0.2
Sevenoaks	5.8	6.4	-0.6	-5.6	2.8	3.1	-0.3	3.0	3.3	-0.3
Shepway	4.9	4.2	0.6	6.7	2.4	2.1	0.3	2.4	2.1	0.3
Swale	5.9	4.6	1.3	10.8	2.9	2.3	0.7	2.9	2.3	0.6
Thanet	5.7	4.9	0.8	6.2	2.9	2.5	0.3	2.8	2.3	0.5
Tonbridge and Malling	6.3	5.7	0.6	5.6	3.2	2.9	0.3	3.2	2.9	0.3
Tunbridge Wells	5.6	6.1	-0.5	-4.8	2.7	2.9	-0.3	3.0	3.2	-0.2
Oxfordshire	**29.8**	**31.1**	**-1.2**	**-2.0**	**14.6**	**15.1**	**-0.5**	**15.2**	**16.0**	**-0.7**
Cherwell	7.3	7.0	0.3	2.3	3.6	3.4	0.2	3.8	3.7	0.1
Oxford	12.0	13.8	-1.8	-13.6	6.0	6.8	-0.8	6.0	7.0	-1.0
South Oxfordshire	7.9	7.9	0.0	-0.1	3.9	3.8	0.1	4.0	4.1	-0.1
Vale of White Horse	6.8	6.5	0.3	2.6	3.2	3.1	0.1	3.5	3.3	0.2
West Oxfordshire	4.6	4.6	0.0	-0.1	2.2	2.2	0.0	2.4	2.4	0.0

Note: Bold figures exclude moves between local authorities within each county and moves between counties and unitary authorities within each Government Office Region.
These statistics may be revised in light of the 2001 Census results.

1 Based on patient register data and patient re-registration recorded in the NHSCR.
2 Absolute net migration rate per 1,000 resident population calculated using mid-2001 population estimates based on 2001 Census.

United Kingdom by countries and, within England and Wales Government Office Regions, counties/unitary authorities, districts and London boroughs

thousands

Age												Area
0-15			16-29			30-59			60 and over			
In	Out	Net inflow	In	Out	Net inflow	In	Out	Net inflow	In	Out	Net inflow	
35.6	**32.4**	**3.2**	**88.9**	**91.1**	**-2.2**	**78.5**	**72.1**	**6.5**	**17.4**	**17.3**	**0.1**	**SOUTH EAST**
1.2	1.4	-0.2	2.1	2.2	-0.1	2.4	2.9	-0.5	0.4	0.4	0.0	**Bracknell Forest UA**
1.4	1.8	-0.4	8.8	6.8	2.1	5.4	5.7	-0.3	0.7	1.3	-0.6	**Brighton and Hove UA**
1.1	0.6	0.5	1.2	1.5	-0.3	2.3	1.3	1.1	0.8	0.6	0.3	**Isle of Wight UA**
2.2	2.1	0.1	3.3	3.2	0.2	3.6	3.8	-0.1	0.7	0.8	-0.1	**Medway UA**
2.1	1.9	0.2	3.9	3.2	0.6	4.2	3.7	0.5	0.8	0.6	0.1	**Milton Keynes UA**
1.4	1.6	-0.2	4.9	4.5	0.4	2.8	3.1	-0.4	0.5	0.6	-0.1	**Portsmouth UA**
1.0	1.6	-0.6	5.9	4.7	1.2	2.8	3.7	-0.9	0.3	0.6	-0.3	**Reading UA**
0.8	1.3	-0.5	2.4	2.2	0.2	2.0	2.7	-0.8	0.2	0.5	-0.2	**Slough UA**
1.0	1.5	-0.5	8.7	7.5	1.2	3.0	3.6	-0.6	0.4	0.7	-0.3	**Southampton UA**
1.4	1.5	-0.1	2.8	3.1	-0.2	3.1	3.2	-0.1	0.5	0.7	-0.2	**West Berkshire UA**
1.5	1.4	0.0	2.9	3.2	-0.3	3.0	3.6	-0.6	0.4	0.7	-0.2	**Windsor and Maidenhead UA**
1.5	1.5	0.0	4.1	3.9	0.2	3.2	3.7	-0.4	0.6	0.7	-0.1	**Wokingham UA**
3.7	**3.6**	**0.2**	**7.5**	**8.7**	**-1.2**	**8.4**	**8.6**	**-0.2**	**1.4**	**1.8**	**-0.4**	**Buckinghamshire**
1.7	1.5	0.1	2.5	2.8	-0.3	3.5	3.2	0.3	0.6	0.6	0.0	Aylesbury Vale
0.8	0.7	0.2	1.6	2.1	-0.5	2.0	1.9	0.1	0.4	0.5	-0.1	Chiltern
0.8	0.8	0.0	1.2	1.3	0.0	1.7	1.7	0.0	0.4	0.4	-0.1	South Bucks
1.3	1.5	-0.2	3.2	3.5	-0.3	2.9	3.5	-0.6	0.4	0.7	-0.3	Wycombe
3.7	**2.6**	**1.1**	**5.4**	**6.3**	**-0.9**	**8.6**	**5.4**	**3.1**	**3.6**	**2.2**	**1.4**	**East Sussex**
0.9	0.7	0.2	1.8	1.7	0.0	2.2	1.5	0.7	1.1	0.6	0.5	Eastbourne
0.9	0.8	0.1	1.2	1.3	0.0	1.8	1.4	0.4	0.6	0.4	0.2	Hastings
0.9	0.8	0.2	1.3	1.5	-0.2	2.0	1.6	0.5	0.9	0.6	0.2	Lewes
1.0	0.8	0.2	1.1	1.4	-0.3	2.1	1.5	0.6	1.1	0.8	0.3	Rother
1.7	1.2	0.5	1.9	2.3	-0.4	3.5	2.6	0.9	1.2	1.1	0.1	Wealden
9.9	**9.1**	**0.8**	**16.6**	**18.4**	**-1.8**	**19.1**	**17.9**	**1.3**	**4.8**	**4.2**	**0.5**	**Hampshire**
1.2	1.4	-0.1	2.4	2.5	-0.1	2.7	3.0	-0.3	0.4	0.6	-0.2	Basingstoke and Deane
1.5	1.3	0.2	1.8	2.2	-0.3	2.8	2.4	0.3	0.7	0.6	0.1	East Hampshire
1.1	1.0	0.2	2.0	2.0	0.0	2.5	2.4	0.2	0.5	0.5	0.0	Eastleigh
1.2	1.0	0.2	1.6	1.7	0.0	2.4	2.0	0.4	0.6	0.5	0.1	Fareham
0.8	1.0	-0.2	1.0	1.1	-0.1	1.2	1.4	-0.1	0.4	0.3	0.1	Gosport
1.2	1.0	0.2	1.7	1.8	-0.1	2.3	2.2	0.1	0.4	0.5	-0.1	Hart
1.1	1.1	0.0	1.5	1.9	-0.4	2.3	2.1	0.1	0.6	0.6	0.0	Havant
1.6	1.2	0.3	2.0	2.5	-0.5	3.4	2.6	0.8	1.3	0.9	0.4	New Forest
1.1	1.7	-0.6	2.2	2.0	0.1	1.8	2.5	-0.7	0.3	0.4	-0.1	Rushmoor
1.4	1.1	0.2	1.8	1.8	0.0	2.5	2.3	0.3	0.5	0.5	0.0	Test Valley
1.4	1.0	0.4	3.0	3.4	-0.4	2.6	2.5	0.2	0.6	0.6	0.0	Winchester
8.9	**6.8**	**2.1**	**15.1**	**17.3**	**-2.2**	**17.5**	**13.4**	**4.1**	**4.6**	**3.7**	**0.8**	**Kent**
1.5	1.0	0.5	1.7	1.6	0.1	2.7	1.9	0.9	0.6	0.5	0.1	Ashford
1.4	1.1	0.3	4.1	4.4	-0.3	2.8	2.3	0.5	0.9	0.7	0.1	Canterbury
0.8	0.9	-0.2	1.6	1.5	0.1	1.7	1.9	-0.2	0.4	0.4	0.0	Dartford
1.1	0.9	0.2	1.2	1.5	-0.3	1.8	1.5	0.4	0.6	0.4	0.2	Dover
0.7	0.7	0.0	1.2	1.2	-0.1	1.4	1.4	0.0	0.3	0.4	-0.1	Gravesham
1.4	1.3	0.1	2.4	2.4	-0.1	3.0	2.6	0.4	0.6	0.6	0.0	Maidstone
1.1	1.1	0.0	1.7	2.1	-0.4	2.5	2.5	0.0	0.5	0.7	-0.2	Sevenoaks
0.9	0.9	0.0	1.2	1.4	-0.2	2.0	1.5	0.5	0.8	0.5	0.3	Shepway
1.4	1.0	0.4	1.6	1.5	0.0	2.4	1.7	0.7	0.5	0.4	0.1	Swale
1.3	1.0	0.3	1.3	1.8	-0.5	2.1	1.5	0.6	0.9	0.6	0.3	Thanet
1.3	1.0	0.3	1.8	1.9	-0.2	2.7	2.2	0.5	0.6	0.6	0.0	Tonbridge and Malling
1.2	1.1	0.1	1.8	2.2	-0.4	2.2	2.3	-0.1	0.5	0.6	-0.1	Tunbridge Wells
4.5	**4.6**	**-0.1**	**14.2**	**14.5**	**-0.3**	**9.5**	**10.1**	**-0.6**	**1.6**	**1.9**	**-0.2**	**Oxfordshire**
1.3	1.5	-0.1	2.5	2.3	0.2	3.0	2.8	0.2	0.6	0.5	0.1	Cherwell
0.9	1.3	-0.4	8.4	8.2	0.2	2.4	3.8	-1.4	0.3	0.5	-0.3	Oxford
1.4	1.3	0.1	2.6	2.8	-0.2	3.3	3.1	0.2	0.6	0.7	-0.1	South Oxfordshire
1.6	1.2	0.4	2.2	2.5	-0.3	2.6	2.3	0.3	0.4	0.5	-0.1	Vale of White Horse
0.8	0.9	0.0	1.4	1.6	-0.2	1.9	1.8	0.1	0.4	0.4	0.1	West Oxfordshire

Table 5.1b - *continued*

Area	Persons In	Persons Out	Persons Net inflow	Persons Rate[2]	Males In	Males Out	Males Net inflow	Females In	Females Out	Females Net inflow
Surrey	**49.7**	**51.3**	**-1.6**	**-1.6**	**23.7**	**24.5**	**-0.9**	**26.0**	**26.8**	**-0.8**
Elmbridge	7.4	7.4	0.0	-0.3	3.6	3.5	0.1	3.7	3.9	-0.2
Epsom and Ewell	4.5	4.5	0.0	0.2	2.1	2.1	0.0	2.3	2.3	0.0
Guildford	9.2	9.5	-0.3	-2.3	4.4	4.5	-0.1	4.8	5.0	-0.2
Mole Valley	4.7	4.9	-0.2	-1.9	2.3	2.3	-0.1	2.4	2.5	-0.1
Reigate and Banstead	7.6	7.1	0.4	3.3	3.5	3.4	0.1	4.0	3.7	0.3
Runnymede	5.7	6.1	-0.3	-4.0	2.7	2.9	-0.2	3.1	3.2	-0.1
Spelthorne	4.6	5.0	-0.4	-4.6	2.2	2.4	-0.3	2.4	2.6	-0.2
Surrey Heath	5.0	5.4	-0.5	-6.1	2.3	2.6	-0.3	2.6	2.8	-0.2
Tandridge	4.9	4.7	0.1	1.5	2.3	2.2	0.0	2.6	2.5	0.1
Waverley	8.0	8.1	-0.2	-1.5	3.8	4.0	-0.1	4.1	4.2	-0.1
Woking	5.2	5.5	-0.3	-3.4	2.6	2.7	-0.1	2.6	2.8	-0.2
West Sussex	**30.2**	**26.8**	**3.4**	**4.5**	**14.5**	**12.9**	**1.7**	**15.6**	**13.9**	**1.7**
Adur	3.3	2.8	0.4	7.3	1.6	1.5	0.1	1.7	1.4	0.3
Arun	7.9	5.9	1.9	13.8	3.8	2.8	1.0	4.1	3.1	0.9
Chichester	6.6	6.1	0.5	4.5	3.1	2.9	0.2	3.4	3.2	0.2
Crawley	4.3	5.1	-0.7	-7.1	2.1	2.5	-0.3	2.2	2.6	-0.4
Horsham	6.8	6.1	0.7	5.6	3.3	3.0	0.3	3.5	3.1	0.4
Mid Sussex	6.2	6.5	-0.3	-2.1	3.0	3.1	-0.1	3.2	3.3	-0.2
Worthing	5.2	4.4	0.8	8.1	2.5	2.1	0.4	2.7	2.3	0.4
SOUTH WEST	**140.2**	**110.4**	**29.8**	**6.0**	**67.8**	**53.0**	**14.8**	**72.4**	**57.5**	**15.0**
Bath and North East Somerset UA	9.6	9.3	0.3	1.6	4.5	4.4	0.2	5.0	4.9	0.1
Bournemouth UA	12.1	11.3	0.8	4.7	6.1	5.6	0.5	6.0	5.7	0.3
Bristol, City of UA	21.1	22.1	-1.0	-2.7	10.4	10.5	-0.1	10.7	11.6	-0.9
North Somerset UA	8.7	7.1	1.6	8.6	4.3	3.5	0.8	4.4	3.6	0.9
Plymouth UA	11.0	10.4	0.6	2.3	5.5	5.2	0.3	5.5	5.2	0.3
Poole UA	8.4	8.0	0.4	3.1	4.1	3.9	0.2	4.3	4.1	0.2
South Gloucestershire UA	11.3	10.4	1.0	3.9	5.5	5.1	0.4	5.9	5.3	0.6
Swindon UA	6.3	6.7	-0.4	-2.0	3.2	3.3	-0.1	3.1	3.3	-0.2
Torbay UA	7.9	5.8	2.1	16.5	4.0	2.9	1.1	3.9	2.9	1.0
Cornwall and Isles of Scilly	**21.2**	**15.9**	**5.3**	**10.5**	**10.3**	**7.7**	**2.5**	**10.9**	**8.2**	**2.7**
Caradon	4.0	3.6	0.4	5.6	2.0	1.7	0.3	2.1	1.9	0.2
Carrick	5.8	4.6	1.2	13.1	2.8	2.2	0.6	3.0	2.4	0.6
Kerrier	4.9	4.2	0.7	7.2	2.4	2.1	0.3	2.5	2.2	0.4
North Cornwall	4.8	3.6	1.2	15.1	2.3	1.7	0.6	2.5	1.8	0.6
Penwith	3.3	2.8	0.5	8.6	1.6	1.3	0.3	1.7	1.4	0.3
Restormel	5.3	4.1	1.2	13.0	2.6	2.0	0.6	2.7	2.0	0.7
Isles of Scilly	0.3	0.2	0.0	6.1	0.1	0.1	0.0	0.1	0.1	0.0
Devon	**32.9**	**25.8**	**7.2**	**10.2**	**15.8**	**12.2**	**3.5**	**17.2**	**13.5**	**3.6**
East Devon	7.6	5.6	2.0	16.1	3.5	2.6	1.0	4.1	3.1	1.1
Exeter	8.0	7.6	0.5	4.2	3.9	3.6	0.3	4.1	4.0	0.1
Mid Devon	4.2	3.6	0.6	8.9	2.1	1.8	0.3	2.1	1.8	0.3
North Devon	4.8	3.9	0.9	10.4	2.3	1.9	0.4	2.5	2.0	0.5
South Hams	5.3	4.8	0.5	6.6	2.5	2.3	0.2	2.8	2.5	0.3
Teignbridge	7.1	5.8	1.2	10.2	3.5	2.8	0.6	3.6	3.0	0.6
Torridge	3.7	2.8	1.0	16.3	1.8	1.3	0.4	1.9	1.4	0.5
West Devon	3.0	2.6	0.4	8.4	1.4	1.3	0.2	1.6	1.4	0.2
Dorset	**20.8**	**16.7**	**4.1**	**10.5**	**10.1**	**7.9**	**2.2**	**10.7**	**8.8**	**1.9**
Christchurch	3.1	2.5	0.5	11.5	1.5	1.2	0.3	1.6	1.4	0.2
East Dorset	5.6	4.5	1.0	12.3	2.7	2.2	0.5	2.9	2.4	0.5
North Dorset	4.0	3.5	0.5	8.9	2.0	1.6	0.3	2.1	1.9	0.2
Purbeck	2.5	2.3	0.2	3.5	1.2	1.1	0.1	1.3	1.2	0.1
West Dorset	5.8	4.7	1.2	12.8	2.9	2.2	0.6	3.0	2.4	0.6
Weymouth and Portland	3.5	2.8	0.7	10.7	1.7	1.4	0.3	1.8	1.4	0.4
Gloucestershire	**20.5**	**18.8**	**1.7**	**3.0**	**9.6**	**9.0**	**0.6**	**10.8**	**9.7**	**1.1**
Cheltenham	6.7	6.7	0.1	0.5	3.1	3.1	0.0	3.6	3.5	0.1
Cotswold	4.9	4.6	0.3	3.4	2.3	2.2	0.1	2.6	2.4	0.2
Forest of Dean	3.7	3.3	0.4	5.1	1.7	1.6	0.1	2.0	1.8	0.3
Gloucester	4.9	5.0	-0.2	-1.7	2.4	2.6	-0.2	2.5	2.5	0.0
Stroud	5.0	4.3	0.7	6.3	2.5	2.1	0.3	2.5	2.2	0.3
Tewkesbury	4.9	4.4	0.5	6.5	2.3	2.1	0.2	2.5	2.3	0.3

Notes: Bold figures exclude moves between local authorities within each county and moves between counties and unitary authorities within each Government Office Region.
These statistics may be revised in light of the 2001 Census results.

1 Based on patient register data and patient re-registration recorded in the NHSCR.
2 Absolute net migration rate per 1,000 resident population calculated using mid-2001 population estimates based on 2001 Census.

United Kingdom by countries and, within England and Wales Government Office Regions, counties/unitary authorities, districts and London boroughs

thousands

Age 0-15 In	Out	Net inflow	16-29 In	Out	Net inflow	30-59 In	Out	Net inflow	60 and over In	Out	Net inflow	Area
7.6	**7.4**	**0.1**	**19.8**	**20.0**	**-0.2**	**18.9**	**19.2**	**-0.3**	**3.4**	**4.7**	**-1.2**	**Surrey**
1.2	1.3	-0.1	2.4	2.2	0.1	3.2	3.1	0.1	0.6	0.8	-0.2	Elmbridge
0.8	0.7	0.1	1.5	1.5	-0.1	1.8	1.8	0.1	0.4	0.5	-0.1	Epsom and Ewell
1.1	1.3	-0.1	4.7	4.2	0.5	2.9	3.3	-0.4	0.5	0.8	-0.3	Guildford
0.8	0.7	0.1	1.4	1.6	-0.2	2.0	1.9	0.1	0.5	0.6	-0.1	Mole Valley
1.3	1.1	0.1	2.5	2.5	0.0	3.1	2.8	0.3	0.7	0.7	0.0	Reigate and Banstead
0.7	0.9	-0.2	2.7	2.5	0.2	2.0	2.2	-0.2	0.4	0.5	-0.1	Runnymede
0.7	0.7	-0.1	1.6	1.6	0.0	1.9	2.1	-0.2	0.4	0.5	-0.1	Spelthorne
1.0	0.9	0.0	1.6	1.8	-0.2	2.0	2.2	-0.2	0.4	0.5	-0.1	Surrey Heath
0.9	0.7	0.2	1.4	1.7	-0.3	2.1	1.9	0.2	0.4	0.5	0.0	Tandridge
1.4	1.2	0.2	2.8	3.4	-0.6	3.0	2.8	0.1	0.7	0.7	0.0	Waverley
0.8	1.0	-0.2	2.0	1.7	0.3	2.1	2.3	-0.2	0.4	0.5	-0.1	Woking
5.1	**4.3**	**0.8**	**9.2**	**9.9**	**-0.7**	**11.8**	**9.5**	**2.3**	**4.0**	**3.1**	**0.9**	**West Sussex**
0.7	0.5	0.2	0.9	1.0	-0.1	1.3	1.0	0.3	0.4	0.4	0.0	Adur
1.2	0.9	0.3	1.9	2.0	-0.1	3.1	2.1	1.1	1.6	0.9	0.7	Arun
1.1	1.1	0.0	1.9	2.1	-0.2	2.6	2.1	0.5	1.0	0.7	0.2	Chichester
0.6	0.9	-0.3	1.9	1.6	0.2	1.6	2.1	-0.5	0.3	0.4	-0.2	Crawley
1.3	1.1	0.3	1.9	1.9	0.0	2.8	2.4	0.4	0.7	0.7	0.0	Horsham
1.2	1.0	0.2	1.9	2.3	-0.5	2.5	2.4	0.1	0.6	0.7	-0.1	Mid Sussex
0.8	0.7	0.1	1.5	1.5	0.0	2.1	1.6	0.5	0.9	0.6	0.2	Worthing
23.5	**15.6**	**7.9**	**51.1**	**51.8**	**-0.7**	**50.6**	**33.8**	**16.8**	**15.0**	**9.3**	**5.7**	**SOUTH WEST**
1.2	1.1	0.2	4.9	4.5	0.4	2.9	3.0	-0.1	0.5	0.7	-0.2	**Bath and North East Somerset UA**
1.3	1.4	-0.1	5.8	5.0	0.8	3.8	3.8	0.0	1.2	1.1	0.1	**Bournemouth UA**
1.8	3.0	-1.2	13.3	10.1	3.2	5.3	7.5	-2.2	0.6	1.4	-0.8	**Bristol, City of UA**
1.5	1.1	0.3	2.4	2.6	-0.2	3.7	2.6	1.0	1.2	0.7	0.5	**North Somerset UA**
1.8	1.6	0.1	5.3	5.0	0.3	3.2	3.1	0.1	0.7	0.7	0.0	**Plymouth UA**
1.5	1.3	0.2	2.7	2.8	-0.1	3.2	3.0	0.2	1.0	0.9	0.2	**Poole UA**
2.0	1.7	0.3	4.2	3.9	0.3	4.3	3.9	0.4	0.8	0.8	0.0	**South Gloucestershire UA**
1.0	1.2	-0.3	2.7	2.3	0.4	2.3	2.7	-0.4	0.4	0.5	-0.1	**Swindon UA**
1.4	0.9	0.5	2.0	2.0	0.0	3.3	2.1	1.2	1.2	0.8	0.3	**Torbay UA**
4.0	**2.5**	**1.6**	**5.4**	**6.5**	**-1.1**	**9.0**	**5.3**	**3.7**	**2.9**	**1.7**	**1.2**	**Cornwall and Isles of Scilly**
0.8	0.6	0.2	0.9	1.2	-0.3	1.8	1.3	0.4	0.6	0.4	0.1	Caradon
1.1	0.7	0.4	1.9	1.9	0.0	2.2	1.6	0.6	0.7	0.5	0.2	Carrick
1.0	0.8	0.2	1.2	1.4	-0.3	2.1	1.5	0.6	0.6	0.5	0.1	Kerrier
0.9	0.6	0.3	1.1	1.2	-0.1	2.0	1.2	0.8	0.8	0.5	0.3	North Cornwall
0.6	0.4	0.1	0.8	1.0	-0.3	1.4	1.0	0.4	0.5	0.3	0.2	Penwith
1.0	0.6	0.3	1.3	1.5	-0.1	2.3	1.5	0.8	0.7	0.5	0.2	Restormel
0.0	0.0	0.0	0.2	0.2	0.0	0.1	0.1	0.0	0.0	0.0	0.0	Isles of Scilly
5.7	**3.7**	**2.0**	**9.7**	**10.8**	**-1.1**	**12.9**	**8.3**	**4.6**	**4.6**	**3.0**	**1.7**	**Devon**
1.3	0.8	0.6	1.8	2.1	-0.3	3.0	1.8	1.2	1.5	0.9	0.5	East Devon
0.8	0.9	-0.1	4.5	3.8	0.7	2.2	2.3	-0.2	0.5	0.4	0.0	Exeter
0.9	0.6	0.3	1.0	1.2	-0.2	1.9	1.4	0.5	0.5	0.4	0.1	Mid Devon
1.1	0.7	0.4	1.1	1.4	-0.3	1.9	1.4	0.6	0.7	0.4	0.2	North Devon
1.0	0.8	0.2	1.3	1.6	-0.3	2.3	1.8	0.5	0.7	0.6	0.1	South Hams
1.3	0.9	0.3	1.6	2.0	-0.4	3.1	2.1	1.0	1.1	0.8	0.3	Teignbridge
0.7	0.5	0.3	0.7	0.9	-0.2	1.6	1.0	0.7	0.7	0.4	0.3	Torridge
0.6	0.5	0.1	0.6	0.8	-0.2	1.3	1.0	0.4	0.4	0.3	0.1	West Devon
4.1	**2.7**	**1.3**	**4.7**	**5.9**	**-1.1**	**8.5**	**5.7**	**2.8**	**3.5**	**2.4**	**1.1**	**Dorset**
0.5	0.4	0.1	0.7	0.7	0.0	1.2	0.9	0.3	0.6	0.5	0.1	Christchurch
1.1	0.7	0.4	1.1	1.4	-0.3	2.4	1.6	0.8	1.0	0.8	0.2	East Dorset
1.1	0.8	0.2	0.9	1.1	-0.2	1.5	1.1	0.4	0.6	0.4	0.1	North Dorset
0.5	0.4	0.1	0.6	0.8	-0.2	1.1	0.9	0.2	0.3	0.3	0.0	Purbeck
1.1	0.7	0.4	1.3	1.6	-0.3	2.4	1.7	0.7	1.1	0.6	0.4	West Dorset
0.6	0.4	0.1	1.0	1.1	-0.1	1.4	1.0	0.5	0.5	0.3	0.2	Weymouth and Portland
3.4	**2.9**	**0.5**	**7.8**	**8.0**	**-0.2**	**7.4**	**6.3**	**1.1**	**1.9**	**1.6**	**0.3**	**Gloucestershire**
0.9	0.8	0.1	3.5	3.3	0.2	1.9	2.1	-0.2	0.5	0.4	0.0	Cheltenham
0.9	0.7	0.1	1.4	1.6	-0.2	2.1	1.7	0.3	0.6	0.5	0.0	Cotswold
0.7	0.7	0.1	1.2	1.2	0.0	1.5	1.2	0.3	0.4	0.3	0.1	Forest of Dean
0.9	1.1	-0.2	1.8	1.6	0.2	1.8	2.0	-0.2	0.4	0.4	0.0	Gloucester
1.0	0.7	0.3	1.4	1.7	-0.3	2.2	1.5	0.7	0.4	0.4	0.0	Stroud
1.0	0.8	0.1	1.4	1.4	0.0	2.0	1.7	0.3	0.5	0.4	0.1	Tewkesbury

Table 5.1b - *continued*

Area	Sex									
	Persons				Males			Females		
	In	Out	Net inflow	Rate[2]	In	Out	Net inflow	In	Out	Net inflow
Somerset	**21.6**	**17.6**	**4.0**	**8.0**	**10.6**	**8.8**	**1.8**	**11.0**	**8.9**	**2.2**
Mendip	5.3	4.8	0.5	5.0	2.6	2.4	0.2	2.7	2.4	0.3
Sedgemoor	5.5	4.5	1.0	9.6	2.7	2.2	0.5	2.7	2.2	0.5
South Somerset	6.9	6.4	0.6	3.7	3.3	3.1	0.2	3.6	3.3	0.4
Taunton Deane	5.9	4.3	1.7	16.5	2.9	2.1	0.8	3.1	2.2	0.9
West Somerset	2.5	2.3	0.2	6.2	1.2	1.1	0.1	1.2	1.1	0.1
Wiltshire	**19.6**	**17.5**	**2.1**	**4.8**	**9.1**	**8.2**	**0.9**	**10.5**	**9.3**	**1.2**
Kennet	4.4	4.3	0.1	0.8	2.1	2.0	0.1	2.3	2.3	0.0
North Wiltshire	6.6	6.7	-0.1	-0.8	3.1	3.2	-0.1	3.4	3.5	0.0
Salisbury	5.7	5.3	0.4	3.1	2.6	2.5	0.1	3.1	2.9	0.3
West Wiltshire	6.0	4.2	1.8	15.1	2.8	2.0	0.8	3.2	2.2	1.0
WALES	**59.3**	**51.8**	**7.5**	**2.6**	**28.6**	**25.1**	**3.5**	**30.7**	**26.6**	**4.1**
Blaenau Gwent	1.2	1.6	-0.4	-5.0	0.6	0.8	-0.2	0.6	0.8	-0.2
Bridgend	4.3	3.5	0.8	6.0	2.0	1.7	0.3	2.2	1.8	0.4
Caerphilly	4.0	3.8	0.2	1.3	1.9	1.9	0.0	2.1	1.9	0.2
Cardiff	15.0	16.0	-1.1	-3.5	6.9	7.4	-0.5	8.1	8.6	-0.5
Carmarthenshire	6.5	5.3	1.2	7.0	3.1	2.6	0.5	3.4	2.7	0.7
Ceredigion	5.5	5.5	0.0	-0.1	2.7	2.6	0.1	2.8	2.9	-0.1
Conwy	6.6	4.8	1.8	16.7	3.2	2.3	0.9	3.4	2.5	0.9
Denbighshire	4.9	4.3	0.6	6.6	2.4	2.1	0.3	2.5	2.2	0.3
Flintshire	5.3	5.0	0.4	2.5	2.5	2.4	0.1	2.8	2.6	0.2
Gwynedd	5.9	5.8	0.1	1.1	2.8	2.7	0.1	3.1	3.0	0.0
Isle of Anglesey	2.6	2.9	-0.3	-4.1	1.2	1.4	-0.2	1.3	1.5	-0.1
Merthyr Tydfil	1.3	1.3	0.0	0.7	0.7	0.6	0.0	0.7	0.6	0.0
Monmouthshire	4.5	3.6	0.9	10.9	2.2	1.8	0.4	2.3	1.8	0.5
Neath Port Talbot	4.0	3.7	0.3	2.1	2.0	1.8	0.1	2.0	1.9	0.2
Newport	4.6	4.3	0.3	2.4	2.3	2.1	0.2	2.3	2.2	0.2
Pembrokeshire	4.4	3.8	0.6	5.2	2.1	1.8	0.3	2.3	2.0	0.3
Powys	5.8	5.0	0.8	6.1	2.7	2.5	0.3	3.0	2.5	0.5
Rhondda, Cynon, Taff	6.1	5.7	0.5	2.1	3.0	2.7	0.3	3.2	3.0	0.2
Swansea	7.7	7.5	0.2	1.0	3.7	3.6	0.1	4.0	3.9	0.1
Torfaen	2.0	2.4	-0.4	-3.9	0.9	1.2	-0.2	1.1	1.2	-0.1
The Vale of Glamorgan	4.8	4.5	0.3	2.5	2.3	2.2	0.1	2.5	2.4	0.1
Wrexham	4.1	3.7	0.5	3.6	2.0	1.9	0.2	2.1	1.8	0.3
SCOTLAND	**54.9**	**51.5**	**3.4**	**0.7**	**26.2**	**25.2**	**1.0**	**28.7**	**26.3**	**2.4**
NORTHERN IRELAND	**11.5**	**11.5**	**0.0**	**0.0**	**5.1**	**5.1**	**0.0**	**6.4**	**6.4**	**0.0**

Note: Bold figures exclude moves between local authorities within each county and moves between counties and unitary authorities within each Government Office Region.
These statistics may be revised in light of the 2001 Census results.

1 Based on patient register data and patient re-registration recorded in the NHSCR.
2 Absolute net migration rate per 1,000 resident population calculated using mid-2001 population estimates based on 2001 Census.

United Kingdom by countries and, within England and Wales Government Office Regions, counties/unitary authorities, districts and London boroughs

thousands

Age 0-15 In	Out	Net inflow	16-29 In	Out	Net inflow	30-59 In	Out	Net inflow	60 and over In	Out	Net inflow	Area
4.6	**2.7**	**1.8**	**5.8**	**7.2**	**-1.4**	**8.5**	**5.9**	**2.6**	**2.8**	**1.8**	**1.0**	**Somerset**
1.2	0.8	0.5	1.3	1.8	-0.5	2.1	1.8	0.3	0.6	0.5	0.2	Mendip
1.2	0.8	0.4	1.2	1.6	-0.4	2.3	1.5	0.8	0.7	0.6	0.1	Sedgemoor
1.5	1.2	0.3	1.7	2.2	-0.5	2.7	2.2	0.5	1.0	0.8	0.3	South Somerset
1.2	0.7	0.5	1.8	1.8	0.0	2.2	1.4	0.8	0.7	0.4	0.4	Taunton Deane
0.3	0.3	0.1	0.9	1.0	-0.1	0.9	0.7	0.2	0.4	0.3	0.1	West Somerset
4.3	**3.5**	**0.8**	**5.3**	**6.3**	**-0.9**	**8.1**	**6.3**	**1.7**	**1.9**	**1.4**	**0.5**	**Wiltshire**
1.0	0.9	0.1	1.2	1.5	-0.4	1.8	1.5	0.3	0.4	0.4	0.0	Kennet
1.5	1.5	0.0	1.8	2.1	-0.3	2.8	2.6	0.2	0.5	0.5	0.1	North Wiltshire
1.2	1.1	0.1	1.6	1.8	-0.2	2.3	2.0	0.4	0.6	0.5	0.1	Salisbury
1.3	0.7	0.5	1.5	1.5	0.0	2.5	1.6	0.9	0.7	0.4	0.3	West Wiltshire
9.9	**7.3**	**2.6**	**23.0**	**26.1**	**-3.1**	**20.4**	**14.4**	**6.0**	**6.0**	**4.0**	**2.0**	**WALES**
0.3	0.4	-0.1	0.4	0.5	-0.1	0.4	0.5	-0.1	0.1	0.1	0.0	Blaenau Gwent
0.9	0.7	0.3	1.1	1.3	-0.2	1.7	1.2	0.5	0.5	0.3	0.2	Bridgend
0.9	0.8	0.2	1.3	1.4	-0.1	1.5	1.4	0.2	0.3	0.3	0.0	Caerphilly
1.4	2.1	-0.6	9.6	8.2	1.4	3.4	4.8	-1.4	0.5	1.0	-0.4	Cardiff
1.3	0.8	0.5	1.8	2.2	-0.4	2.6	1.7	0.9	0.8	0.6	0.3	Carmarthenshire
0.6	0.6	0.0	2.9	3.2	-0.4	1.5	1.2	0.3	0.5	0.4	0.1	Ceredigion
1.4	0.9	0.5	1.4	1.6	-0.2	2.6	1.6	1.0	1.2	0.7	0.5	Conwy
1.0	0.8	0.2	1.3	1.5	-0.2	1.9	1.4	0.5	0.7	0.6	0.1	Denbighshire
1.1	0.9	0.1	1.6	1.8	-0.2	2.2	1.8	0.4	0.5	0.5	0.1	Flintshire
0.9	0.9	0.0	2.4	2.6	-0.2	1.9	1.7	0.2	0.7	0.6	0.1	Gwynedd
0.5	0.5	0.0	0.6	0.9	-0.3	1.0	1.0	0.1	0.4	0.5	0.0	Isle of Anglesey
0.3	0.3	0.0	0.4	0.4	0.0	0.5	0.5	0.0	0.1	0.1	0.0	Merthyr Tydfil
0.9	0.6	0.4	1.1	1.3	-0.2	2.0	1.3	0.7	0.5	0.4	0.1	Monmouthshire
1.0	0.7	0.2	1.2	1.3	-0.2	1.5	1.3	0.2	0.4	0.4	0.0	Neath Port Talbot
1.0	0.8	0.1	1.6	1.6	0.0	1.7	1.5	0.3	0.3	0.4	-0.1	Newport
0.9	0.8	0.1	1.1	1.5	-0.4	1.7	1.2	0.6	0.7	0.4	0.3	Pembrokeshire
1.2	0.9	0.3	1.2	1.9	-0.6	2.4	1.7	0.8	1.0	0.6	0.3	Powys
1.1	1.0	0.1	2.2	2.6	-0.3	2.2	1.7	0.5	0.6	0.4	0.2	Rhondda, Cynon, Taff
1.0	1.0	0.1	4.1	4.1	0.0	2.1	2.0	0.1	0.4	0.4	0.0	Swansea
0.5	0.5	-0.1	0.6	0.8	-0.2	0.7	0.8	-0.1	0.2	0.2	0.0	Torfaen
1.0	0.9	0.1	1.4	1.6	-0.2	1.9	1.6	0.3	0.5	0.4	0.1	The Vale of Glamorgan
0.7	0.6	0.1	1.5	1.5	0.0	1.5	1.2	0.3	0.4	0.3	0.1	Wrexham
10.0	**8.5**	**1.4**	**21.1**	**22.8**	**-1.8**	**20.2**	**17.5**	**2.7**	**3.7**	**2.7**	**1.0**	**SCOTLAND**
2.7	**2.2**	**0.4**	**4.4**	**6.2**	**-1.8**	**3.9**	**2.7**	**1.2**	**0.6**	**0.4**	**0.2**	**NORTHERN IRELAND**

Table 5.2a Movements[1] between countries of the United Kingdom and, within England, metropolitan counties and remainders of Government Office Regions in England with the rest of the United Kingdom during the year ending June 2001: flows within and between areas

Area of destination		Area of origin: North East – Tyne and Wear Met County	North East – Remainder	North West – Greater Manchester Met County	North West – Merseyside Met County	North West – Remainder	Yorkshire and The Humber – South Yorkshire Met County	Yorkshire and The Humber – West Yorkshire Met County	Yorkshire and The Humber – Remainder	East Midlands
North East	Tyne and Wear (Met County)	**7.0**	7.2	0.9	0.4	1.6	0.5	1.2	1.2	1.2
	Remainder	8.8	**4.3**	1.0	0.5	1.9	0.6	1.5	3.1	1.8
North West	Greater Manchester (Met County)	1.0	1.0	**37.7**	3.4	12.1	1.5	3.5	1.6	3.6
	Merseyside (Met County)	0.5	0.5	3.0	**13.1**	6.3	0.5	1.0	0.6	1.2
	Remainder	1.5	2.0	16.4	7.7	**19.8**	1.4	4.3	3.1	4.4
Yorkshire and The Humber	South Yorkshire (Met County)	0.6	0.7	1.4	0.6	1.7	**8.3**	3.2	2.6	6.1
	West Yorkshire (Met County)	1.4	1.7	3.9	1.3	4.1	3.4	**18.3**	7.4	4.3
	Remainder	1.3	3.6	1.7	0.7	3.3	3.4	10.3	**7.0**	5.9
East Midlands		1.4	2.0	4.1	1.4	4.6	7.5	4.3	5.9	**39.4**
West Midlands	West Midlands (Met County)	0.7	0.6	1.8	0.7	1.8	1.0	1.6	0.8	5.1
	Remainder	0.5	0.8	2.1	1.0	4.9	0.9	1.5	1.3	8.2
East		1.4	1.9	2.7	1.5	3.6	1.9	2.9	2.9	13.1
London		2.7	2.3	6.2	2.6	4.4	2.5	4.9	3.1	10.5
South East		1.9	2.7	4.5	2.2	6.1	2.3	4.0	4.0	13.7
South West		0.9	1.4	3.3	1.6	4.5	1.6	2.4	2.5	8.6
Wales		0.4	0.6	3.6	2.2	5.4	0.6	1.2	1.0	3.2
Scotland[2]		2.0	2.7	2.4	1.0	4.7	0.9	2.3	2.5	3.6
Northern Ireland[3]		0.2	0.2	0.5	0.3	0.6	0.1	0.3	0.3	0.6
Totals: Excluding moves within areas		27.3	32.0	59.5	29.1	71.4	30.8	50.5	44.0	95.1
Totals: Including moves within areas		34.3	36.3	97.2	42.2	91.2	39.0	68.8	51.0	134.5

Notes: Bold figures represent moves between Health Areas within the areas shown.
These statistics may be revised in light of the 2001 Census results.

1 Based on NHS patient re-registration (see Appendix A).
2 Scotland data for within area moves provided by GRO Scotland.
3 Northern Ireland data for within area moves not available.

Table 5.2b Movements[1] between countries of the United Kingdom and, within England, Government Office Regions of England and the rest of the United Kingdom during the year ending June 2001: flows within and between regions

Area of destination	Area of origin: North East	North West	Yorkshire and The Humber	East Midlands	West Midlands	East	London
North East	**27.3**	6.2	8.2	3.1	2.5	3.1	4.1
North West	6.4	**119.5**	17.6	9.1	12.5	7.8	12.1
Yorkshire and The Humber	9.3	18.7	**63.9**	16.3	7.8	8.7	9.4
East Midlands	3.4	10.1	17.8	**39.4**	15.6	17.8	12.5
West Midlands	2.7	12.3	7.2	13.3	**81.3**	7.8	11.3
East	3.3	7.7	7.7	13.1	7.2	**61.1**	59.0
London	5.0	13.2	10.6	10.5	11.9	29.1	**230.5**
South East	4.6	12.8	10.4	13.7	13.9	28.0	86.7
South West	2.4	9.5	6.5	8.6	15.8	13.5	22.1
Wales	1.0	11.1	2.8	3.2	9.7	3.7	5.6
Scotland[2]	4.7	8.1	5.7	3.6	3.4	4.9	7.7
Northern Ireland[3]	0.4	1.4	0.7	0.6	0.7	1.1	1.7
Total: Excluding moves within areas	43.2	111.1	94.9	95.1	101.0	125.4	232.2
Total: Including moves within areas	70.6	230.6	158.8	134.5	182.4	186.5	462.6

Notes: Bold figures represent moves between Health Areas within the areas shown.
These statistics may be revised in light of the 2001 Census results.

1 Based on NHS patient re-registration (see Appendix A).
2 Scotland data for within area moves provided by GRO Scotland.
3 Northern Ireland data for within area moves not available.

thousands

West Midlands: West Midlands Met County	West Midlands: Remainder	East	London	South East	South West	Wales	Scotland[2]	N Ireland[3]	Totals: Excluding moves within areas	Totals: Including moves within areas	Area of destination	
0.5	0.5	1.2	1.9	1.8	0.8	0.3	1.9	0.3	23.5	30.5	Tyne and Wear (Met County)	North East
0.6	0.9	1.9	2.2	2.8	1.5	0.6	2.3	0.2	32.0	36.4	Remainder	
1.9	2.0	2.6	5.4	4.2	2.5	2.6	2.3	0.7	51.9	89.7	Greater Manchester (Met County)	North West
0.9	1.0	1.3	2.1	2.2	1.4	1.7	0.9	0.5	25.5	38.6	Merseyside (Met County)	
2.1	4.6	3.9	4.6	6.8	4.0	4.5	4.6	0.9	76.9	96.7	Remainder	
0.9	1.1	2.0	2.2	2.8	1.5	0.7	0.9	0.1	29.1	37.4	South Yorkshire (Met County)	Yorkshire and The Humber
1.6	1.6	3.1	4.2	4.0	2.1	1.1	1.7	0.4	47.4	65.7	West Yorkshire (Met County)	
1.0	1.6	3.6	3.0	4.8	2.4	1.0	2.2	0.3	50.0	56.9	Remainder	
6.3	9.2	17.8	12.5	19.6	7.9	3.2	3.3	0.7	111.8	151.2		East Midlands
30.1	13.2	3.2	6.8	5.9	4.2	2.3	1.2	0.3	51.4	81.4	West Midlands (Met County)	West Midlands
23.0	**15.1**	4.5	4.5	9.1	8.0	5.5	1.7	0.6	78.3	93.4	Remainder	
3.4	3.9	**61.1**	59.0	28.3	9.8	3.3	4.6	0.8	144.9	206.0		East
7.3	4.6	29.1	**230.5**	52.9	16.1	5.3	7.5	1.4	163.6	394.0		London
6.2	7.7	28.0	86.7	**124.5**	33.1	8.0	7.8	1.4	220.3	344.9		South East
6.2	9.7	13.5	22.1	47.5	**60.3**	9.4	4.3	0.7	140.2	200.5		South West
3.2	6.5	3.7	5.6	10.0	10.1	**22.1**	1.7	0.4	59.3	81.4		Wales
1.4	2.0	4.9	7.7	8.7	4.3	1.9	**80.1**	1.8	54.9	135.0		Scotland[2]
0.4	0.3	1.1	1.7	1.5	0.7	0.3	2.5	-	11.5	11.5		Northern Ireland[3]
66.7	70.5	125.4	232.2	212.9	110.4	51.8	51.5	11.5	**1372.5**	:	Totals: Excluding moves within areas	
96.8	85.6	186.5	462.6	337.4	170.7	73.9	131.6	11.5	:	**2151.1**	Totals: Including moves within areas	

thousands

South East	South West	Wales	Scotland[2]	N Ireland[3]	**Totals**: Excluding moves within regions	**Totals**: Including moves within regions	Area of destination
4.6	2.3	0.9	4.1	0.5	39.5	66.8	North East
13.2	8.0	8.8	7.8	2.1	105.5	225.0	North West
11.6	6.0	2.8	4.8	0.8	96.1	160.0	Yorkshire and The Humber
19.6	7.9	3.2	3.3	0.7	111.8	151.2	East Midlands
15.0	12.2	7.9	2.9	0.9	93.5	174.8	West Midlands
28.3	9.8	3.3	4.6	0.8	144.9	206.0	East
52.9	16.1	5.3	7.5	1.4	163.6	394.0	London
124.5	33.1	8.0	7.8	1.4	220.3	344.9	South East
47.5	**60.3**	9.4	4.3	0.7	140.2	200.5	South West
10.0	10.1	**22.1**	1.7	0.4	59.3	81.4	Wales
8.7	4.3	1.9	**80.1**	1.8	54.9	135.0	Scotland[2]
1.5	0.7	0.3	2.5	-	11.5	11.5	Northern Ireland[3]
212.9	110.4	51.8	51.5	11.5	**1241.1**	:	Total: Excluding moves within areas
337.4	170.7	73.9	131.6	11.5	:	**2151.1**	Total: Including moves within areas

Table 5.3a Internal migration to/from each country/GOR during the year ending June 2001

United Kingdom by countries and, within England, Government Office Regions

thousands

Country/GOR	Into Country/ GOR from England[1]	Out of Country/ GOR to England[1]	Net	Into Country/ GOR from Wales[1]	Out of Country/ GOR to Wales[1]	Net	Into Country/ GOR from Scotland[1]	Out of Country/ GOR to Scotland[1]	Net	Into Country/ GOR from Northern Ireland[1]	Out of Country/ GOR to Northern Ireland[1]	Net
United Kingdom	117.0	106.0	+11.0	51.8	59.3	-7.5	51.5	54.9	-3.4	11.5	11.5	+0.0
England	-	-	-	49.5	57.1	-7.6	47.2	51.2	-3.9	9.2	8.7	+0.6
North East	34.0	37.2	-3.2	0.9	1.0	-0.1	4.1	4.7	-0.6	0.5	0.4	+0.1
North West	86.8	90.5	-3.7	8.8	11.1	-2.4	7.8	8.1	-0.4	2.1	1.4	+0.7
Yorkshire and The Humber	87.7	85.8	+1.9	2.8	2.8	+0.0	4.8	5.7	-0.8	0.8	0.7	+0.1
East Midlands	104.6	87.8	+16.8	3.2	3.2	-0.1	3.3	3.6	-0.2	0.7	0.6	+0.1
West Midlands	81.8	87.2	-5.4	7.9	9.7	-1.9	2.9	3.4	-0.5	0.9	0.7	+0.2
East	136.2	115.7	+20.5	3.3	3.7	-0.4	4.6	4.9	-0.4	0.8	1.1	-0.3
London	149.4	217.1	-67.8	5.3	5.6	-0.3	7.5	7.7	-0.2	1.4	1.7	-0.3
South East	203.2	192.7	+10.5	8.0	10.0	-2.0	7.8	8.7	-0.9	1.4	1.5	-0.1
South West	125.7	95.4	+30.3	9.4	10.1	-0.6	4.3	4.3	0.0	0.7	0.7	+0.0
Wales	57.1	49.5	+7.6	-	-	-	1.7	1.9	-0.2	0.4	0.3	+0.1
Scotland	51.2	47.2	+3.9	1.9	1.7	+0.2	-	-	-	1.8	2.5	-0.7
Northern Ireland	8.7	9.2	-0.6	0.3	0.4	-0.1	2.5	1.8	+0.7	-	-	-

Note: These statistics may be revised in light of the 2001 Census results.

1 Based on NHS patient re-registrations (see appendix A).

Table 5.3b Total migration to/from each country/GOR during the year ending June 2001

United Kingdom by countries and, within England, Government Office Regions

thousands

Country/GOR	Total internal Migration[1]			Total International Migration[2]			Total Migration[3]		
	Into Country/ GOR from UK	Out of Country/ GOR to UK	Net	In	Out	Net	In	Out	Net
United Kingdom	-	-	-	..	..	..	..	..	..
England	106.0	117.0	-11.0	..	..	..	..	..	..
North East	39.5	43.2	-3.7	..	..	..	..	..	..
North West	105.5	111.1	-5.7	..	..	..	..	..	..
Yorkshire and The Humber	96.1	94.9	+1.2	..	..	..	..	..	..
East Midlands	111.8	95.1	+16.7	..	..	..	..	..	..
West Midlands	93.5	101.0	-7.6	..	..	..	..	..	..
East	144.9	125.4	+19.5	..	..	..	..	..	..
London	163.6	232.2	-68.6	..	..	..	..	..	..
South East	220.3	212.9	+7.5	..	..	..	..	..	..
South West	140.2	110.4	+29.8	..	..	..	..	..	..
Wales	59.3	51.8	+7.5	..	..	..	..	..	..
Scotland	54.9	51.5	+3.4	..	..	..	..	..	..
Northern Ireland	11.5	11.5	+0.0	..	..	..	..	..	..

Note: These statistics may be revised in light of the 2001 Census results.

1 Based on NHS patient re-registrations (see appendix A).
2 For explanation see appendix A.
3 Excludes migration with Channel Islands and Isle of Man.

Table A1 Mid-2001 population estimates: estimated resident population by single year of age and sex

United Kingdom

thousands

Age	Persons	Males	Females	Age	Persons	Males	Females
All ages	**58,836.7**	**28,611.3**	**30,225.3**				
0-4	**3,477.2**	**1,781.0**	**1,696.2**	**50-54**	**4,013.8**	**1,989.5**	**2,024.3**
0	662.4	338.4	324.0	50	739.0	366.1	372.9
1	677.0	347.3	329.6	51	760.7	376.4	384.3
2	697.0	357.1	339.9	52	784.8	389.2	395.7
3	710.5	363.8	346.7	53	840.8	416.8	424.0
4	730.4	374.4	356.1	54	888.4	441.0	447.4
5-9	**3,729.2**	**1,910.5**	**1,818.7**	**55-59**	**3,386.0**	**1,674.9**	**1,711.0**
5	722.8	370.6	352.2	55	722.4	358.1	364.3
6	726.1	372.6	353.5	56	705.7	349.4	356.3
7	744.4	381.2	363.1	57	693.0	343.5	349.5
8	755.5	386.7	368.7	58	664.2	328.4	335.9
9	780.5	399.4	381.1	59	600.5	295.6	305.0
10-14	**3,884.8**	**1,990.2**	**1,894.6**	**60-64**	**2,879.7**	**1,409.3**	**1,470.4**
10	790.1	404.8	385.3	60	563.5	276.4	287.2
11	777.6	398.5	379.0	61	585.5	287.0	298.5
12	771.0	394.9	376.1	62	586.2	286.7	299.5
13	781.8	400.7	381.1	63	579.5	283.8	295.8
14	764.4	391.3	373.1	64	565.0	275.5	289.6
15-19	**3,672.9**	**1,876.8**	**1,796.1**	**65-69**	**2,600.2**	**1,243.5**	**1,356.7**
15	763.4	391.7	371.7	65	549.1	266.1	283.0
16	759.7	389.8	369.9	66	532.0	256.9	275.1
17	731.5	375.2	356.3	67	510.1	244.1	265.9
18	714.6	364.8	349.8	68	504.4	239.1	265.3
19	703.7	355.4	348.3	69	504.7	237.4	267.3
20-24	**3,554.7**	**1,771.4**	**1,783.2**	**70-74**	**2,339.6**	**1,060.3**	**1,279.3**
20	739.1	371.5	367.6	70	502.1	232.7	269.4
21	741.4	371.6	369.8	71	487.3	222.7	264.6
22	714.0	356.1	357.8	72	467.3	211.5	255.8
23	680.1	336.8	343.4	73	445.4	199.7	245.7
24	680.1	335.5	344.7	74	437.5	193.6	243.8
25-29	**3,845.0**	**1,885.9**	**1,959.0**	**75-79**	**1,962.8**	**817.4**	**1,145.4**
25	699.3	343.9	355.4	75	426.2	185.3	240.8
26	727.1	356.6	370.5	76	406.1	173.0	233.1
27	756.5	370.6	385.9	77	389.4	162.1	227.4
28	806.9	396.0	410.9	78	372.0	150.9	221.1
29	855.3	418.8	436.4	79	369.1	146.1	223.0
30-34	**4,486.8**	**2,197.0**	**2,289.8**	**80-84**	**1,326.7**	**488.7**	**838.0**
30	877.7	429.0	448.7	80	365.0	140.6	224.5
31	872.0	427.0	444.9	81	340.3	128.8	211.5
32	898.1	439.2	458.9	82	237.4	86.8	150.6
33	908.1	444.4	463.7	83	191.4	67.1	124.3
34	931.0	457.4	473.6	84	192.6	65.5	127.1
35-39	**4,633.8**	**2,281.4**	**2,352.4**	**85-89**	**752.6**	**227.3**	**525.3**
35	931.6	457.3	474.3	85	183.6	59.7	123.9
36	940.8	462.8	478.0	86	175.2	54.5	120.7
37	935.6	460.4	475.2	87	154.1	45.8	108.3
38	923.0	454.9	468.1	88	131.2	37.6	93.7
39	902.7	446.0	456.7	89	108.5	29.7	78.8
40-44	**4,170.7**	**2,065.8**	**2,104.9**	**90 and over**	**375.7**	**84.7**	**290.9**
40	878.2	434.0	444.2				
41	846.3	419.3	427.0				
42	830.7	411.6	419.2	**Under 16**	**11,854.6**	**6,073.4**	**5,781.2**
43	820.4	406.8	413.6				
44	795.1	394.1	400.9	**Under 18**	**13,345.8**	**6,838.4**	**6,507.5**
45-49	**3,744.5**	**1,855.5**	**1,889.0**	**16-44**	**23,600.5**	**11,686.7**	**11,913.8**
45	767.3	380.6	386.7				
46	747.3	370.3	377.0	**45-64/59**[1]	**12,553.6**	**6,929.3**	**5,624.2**
47	751.7	372.3	379.4				
48	743.4	368.4	375.0	**65/60**[2] **and over**	**10,828.0**	**3,921.9**	**6,906.1**
49	734.9	363.9	371.0				

1 45-64 for males; 45-59 for females
2 65 and over for males; 60 and over for females

Table A2 Mid-2001 population estimates: estimated resident population by single year of age and sex

England and Wales

thousands

Age	Persons	Males	Females
All ages	**52,084.5**	**25,354.7**	**26,729.8**
0-4	**3,086.2**	**1,580.1**	**1,506.1**
0	588.8	300.9	287.9
1	601.6	308.4	293.2
2	618.2	316.7	301.5
3	630.1	322.5	307.6
4	647.5	331.6	315.9
5-9	**3,300.6**	**1,691.2**	**1,609.4**
5	640.8	328.5	312.3
6	642.9	329.8	313.1
7	659.0	337.7	321.3
8	668.1	342.1	325.9
9	689.8	353.0	336.8
10-14	**3,429.4**	**1,756.6**	**1,672.9**
10	698.7	358.0	340.7
11	687.9	352.6	335.3
12	681.3	348.9	332.4
13	689.3	353.1	336.2
14	672.2	344.0	328.2
15-19	**3,225.5**	**1,649.7**	**1,575.8**
15	670.9	344.4	326.6
16	667.1	342.6	324.5
17	643.3	330.3	313.0
18	628.0	320.8	307.2
19	616.2	311.6	304.6
20-24	**3,129.2**	**1,558.5**	**1,570.7**
20	647.9	325.5	322.4
21	650.7	325.9	324.8
22	627.4	312.7	314.7
23	600.2	297.2	303.0
24	602.9	297.1	305.8
25-29	**3,415.7**	**1,676.3**	**1,739.4**
25	618.2	303.8	314.4
26	645.7	316.9	328.7
27	672.7	329.6	343.2
28	717.7	352.5	365.2
29	761.4	373.5	387.9
30-34	**3,978.1**	**1,950.3**	**2,027.8**
30	779.0	381.2	397.9
31	773.7	379.4	394.3
32	796.2	389.9	406.3
33	804.2	394.2	409.9
34	825.0	405.6	419.4
35-39	**4,100.6**	**2,023.1**	**2,077.5**
35	826.7	406.6	420.1
36	832.4	410.2	422.2
37	827.8	408.4	419.4
38	816.0	403.1	412.9
39	797.7	394.8	402.9
40-44	**3,674.0**	**1,823.6**	**1,850.4**
40	774.6	383.7	391.0
41	745.4	370.0	375.5
42	731.4	363.3	368.2
43	722.5	359.1	363.4
44	700.0	347.6	352.4
45-49	**3,303.4**	**1,636.4**	**1,666.9**
45	674.5	334.6	339.9
46	658.2	326.0	332.1
47	663.3	328.3	334.9
48	656.5	325.2	331.3
49	650.9	322.2	328.7

Age	Persons	Males	Females
50-54	**3,564.6**	**1,767.1**	**1,797.5**
50	653.5	323.7	329.8
51	674.2	333.5	340.7
52	696.4	345.3	351.1
53	749.8	371.8	377.9
54	790.7	392.8	397.9
55-59	**3,006.5**	**1,489.1**	**1,517.4**
55	643.8	319.1	324.7
56	629.7	312.1	317.7
57	615.1	305.2	309.9
58	588.0	291.3	296.7
59	529.8	261.3	268.5
60-64	**2,544.5**	**1,249.2**	**1,295.3**
60	497.0	244.5	252.5
61	517.2	254.3	263.0
62	518.2	254.3	263.9
63	512.6	251.6	260.9
64	499.5	244.6	254.9
65-69	**2,295.2**	**1,102.8**	**1,192.5**
65	484.7	235.8	248.8
66	469.5	227.7	241.8
67	449.7	216.2	233.5
68	445.7	212.3	233.4
69	445.7	210.8	234.9
70-74	**2,074.6**	**944.9**	**1,129.6**
70	444.6	207.1	237.5
71	431.6	198.3	233.3
72	414.0	188.3	225.7
73	395.5	178.2	217.2
74	388.9	173.0	215.9
75-79	**1,750.5**	**732.6**	**1,017.9**
75	378.3	165.4	212.9
76	361.6	154.7	206.9
77	347.0	145.1	201.9
78	333.0	135.9	197.2
79	330.6	131.5	199.1
80-84	**1,190.1**	**440.5**	**749.5**
80	328.2	126.9	201.4
81	304.8	115.9	188.9
82	213.1	78.3	134.7
83	170.9	60.3	110.6
84	173.0	59.1	114.0
85-89	**677.1**	**205.9**	**471.2**
85	164.9	54.1	110.8
86	157.5	49.3	108.2
87	138.5	41.5	97.0
88	118.2	34.0	84.2
89	97.9	27.0	70.9
90 and over	**339.0**	**76.8**	**262.3**
Under 16	**10,487.1**	**5,372.2**	**5,114.9**
Under 18	**11,797.5**	**6,045.2**	**5,752.3**
16-44	**20,852.1**	**10,337.1**	**10,515.0**
45-64/59[1]	**11,123.6**	**6,141.9**	**4,981.8**
65/60[2] **and over**	**9,621.7**	**3,503.5**	**6,118.2**

[1] 45-64 for males; 45-59 for females

[2] 65 and over for males; 60 and over for females

Table A3 Mid-2001 population estimates: estimated resident population by single year of age and sex

England
thousands

Age	Persons	Males	Females
All ages	**49,181.3**	**23,950.6**	**25,230.8**
0-4	**2,919.1**	**1,494.5**	**1,424.7**
0	557.2	284.6	272.6
1	569.6	292.0	277.7
2	585.0	299.7	285.3
3	595.6	304.9	290.7
4	611.7	313.2	298.5
5-9	**3,115.7**	**1,596.6**	**1,519.1**
5	605.3	310.1	295.2
6	607.3	311.7	295.6
7	622.2	318.9	303.3
8	630.3	322.9	307.5
9	650.5	333.0	317.5
10-14	**3,233.3**	**1,655.6**	**1,577.8**
10	659.2	337.6	321.6
11	649.0	332.5	316.6
12	642.0	328.6	313.5
13	649.8	332.9	316.9
14	633.3	324.0	309.3
15-19	**3,040.1**	**1,556.0**	**1,484.1**
15	632.3	324.5	307.8
16	628.4	322.8	305.5
17	606.6	311.7	294.9
18	592.1	302.7	289.3
19	580.7	294.2	286.5
20-24	**2,959.6**	**1,474.0**	**1,485.6**
20	610.1	306.9	303.2
21	613.2	307.2	306.0
22	593.6	295.7	297.9
23	569.8	282.2	287.7
24	572.9	282.2	290.7
25-29	**3,251.0**	**1,595.9**	**1,655.1**
25	588.3	288.9	299.4
26	614.8	301.8	313.0
27	640.5	313.9	326.6
28	683.3	335.9	347.4
29	724.1	355.3	368.8
30-34	**3,780.4**	**1,855.0**	**1,925.3**
30	740.6	362.5	378.1
31	735.4	360.9	374.4
32	756.3	370.7	385.7
33	764.1	375.0	389.2
34	784.0	386.0	398.1
35-39	**3,888.5**	**1,919.4**	**1,969.1**
35	785.1	386.3	398.8
36	789.6	389.2	400.4
37	784.8	387.3	397.4
38	773.4	382.4	391.0
39	755.6	374.1	381.6
40-44	**3,477.8**	**1,727.6**	**1,750.2**
40	733.7	363.7	370.0
41	706.1	350.7	355.4
42	692.4	344.1	348.3
43	683.7	340.2	343.5
44	661.9	329.0	333.0
45-49	**3,118.7**	**1,545.5**	**1,573.2**
45	637.8	316.6	321.2
46	622.0	308.1	313.8
47	625.9	310.0	315.9
48	619.2	306.9	312.3
49	613.8	303.9	309.9

Age	Persons	Males	Females
50-54	**3,357.6**	**1,664.6**	**1,693.1**
50	615.7	305.1	310.5
51	634.7	313.9	320.8
52	655.8	325.1	330.6
53	706.0	350.0	355.9
54	745.5	370.4	375.2
55-59	**2,827.4**	**1,400.5**	**1,426.9**
55	606.6	300.7	305.8
56	592.8	293.9	298.9
57	578.7	287.1	291.6
58	552.4	273.7	278.7
59	496.9	245.1	251.9
60-64	**2,391.2**	**1,173.9**	**1,217.3**
60	466.0	229.4	236.7
61	486.2	238.8	247.4
62	487.3	239.1	248.2
63	482.0	236.7	245.3
64	469.7	230.0	239.7
65-69	**2,156.6**	**1,036.3**	**1,120.3**
65	455.6	221.6	233.9
66	440.9	213.8	227.2
67	422.4	203.1	219.2
68	418.7	199.5	219.2
69	419.1	198.3	220.8
70-74	**1,949.0**	**887.7**	**1,061.3**
70	417.9	194.6	223.3
71	405.6	186.3	219.3
72	389.0	176.9	212.1
73	371.7	167.5	204.2
74	364.9	162.4	202.4
75-79	**1,641.0**	**686.8**	**954.2**
75	354.4	154.9	199.5
76	338.7	145.0	193.8
77	325.3	136.0	189.3
78	312.5	127.4	185.1
79	310.1	123.5	186.5
80-84	**1,116.9**	**413.9**	**703.0**
80	308.4	119.3	189.2
81	286.6	109.2	177.4
82	199.5	73.3	126.2
83	159.9	56.5	103.4
84	162.5	55.6	106.9
85-89	**638.0**	**194.3**	**443.7**
85	155.2	51.0	104.2
86	148.4	46.5	101.9
87	130.5	39.2	91.4
88	111.6	32.1	79.5
89	92.3	25.5	66.8
90 and over	**319.4**	**72.5**	**247.0**
Under 16	**9,900.4**	**5,071.2**	**4,829.3**
Under 18	**11,135.5**	**5,705.7**	**5,429.7**
16-44	**19,765.1**	**9,803.5**	**9,961.6**
45-64/59[1]	**10,477.6**	**5,784.5**	**4,693.1**
65/60[2] **and over**	**9,038.2**	**3,291.5**	**5,746.8**

1 45-64 for males; 45-59 for females
2 65 and over for males; 60 and over for females

Table A4 Mid-2001 population estimates: estimated resident population by single year of age and sex

Wales
thousands

Age	Persons	Males	Females	Age	Persons	Males	Females
All ages	**2,903.2**	**1,404.1**	**1,499.1**				
0-4	**167.1**	**85.7**	**81.4**	**50-54**	**207.0**	**102.6**	**104.4**
0	31.6	16.3	15.3	50	37.9	18.6	19.3
1	32.0	16.4	15.6	51	39.5	19.6	19.9
2	33.2	17.0	16.3	52	40.6	20.1	20.5
3	34.5	17.6	16.9	53	43.8	21.8	22.0
4	35.8	18.4	17.4	54	45.2	22.4	22.8
5-9	**184.8**	**94.5**	**90.3**	**55-59**	**179.1**	**88.6**	**90.4**
5	35.4	18.3	17.1	55	37.2	18.4	18.8
6	35.6	18.1	17.4	56	36.9	18.2	18.7
7	36.8	18.8	18.0	57	36.4	18.2	18.3
8	37.7	19.3	18.4	58	35.6	17.6	18.0
9	39.2	20.0	19.3	59	32.9	16.3	16.6
10-14	**196.1**	**101.0**	**95.1**	**60-64**	**153.3**	**75.3**	**77.9**
10	39.5	20.4	19.1	60	31.0	15.1	15.8
11	38.8	20.1	18.7	61	31.0	15.4	15.6
12	39.3	20.3	19.0	62	30.9	15.2	15.7
13	39.5	20.2	19.3	63	30.6	15.0	15.6
14	39.0	20.0	19.0	64	29.8	14.6	15.2
15-19	**185.4**	**93.7**	**91.7**	**65-69**	**138.6**	**66.5**	**72.2**
15	38.6	19.9	18.8	65	29.1	14.2	14.9
16	38.7	19.8	19.0	66	28.6	13.9	14.6
17	36.6	18.6	18.0	67	27.3	13.1	14.2
18	35.9	18.0	17.8	68	27.0	12.8	14.2
19	35.5	17.4	18.1	69	26.6	12.5	14.2
20-24	**169.6**	**84.5**	**85.1**	**70-74**	**125.5**	**57.2**	**68.3**
20	37.8	18.6	19.2	70	26.7	12.5	14.2
21	37.5	18.8	18.8	71	26.0	12.0	14.0
22	33.9	17.1	16.8	72	25.0	11.4	13.6
23	30.4	15.0	15.4	73	23.8	10.7	13.1
24	30.0	14.9	15.0	74	24.0	10.5	13.5
25-29	**164.7**	**80.4**	**84.2**	**75-79**	**109.5**	**45.8**	**63.8**
25	29.9	14.9	15.0	75	23.9	10.5	13.4
26	30.9	15.1	15.8	76	22.8	9.8	13.1
27	32.2	15.6	16.6	77	21.7	9.1	12.6
28	34.5	16.7	17.8	78	20.5	8.4	12.1
29	37.3	18.2	19.1	79	20.6	8.0	12.5
30-34	**197.7**	**95.3**	**102.4**	**80-84**	**73.1**	**26.6**	**46.5**
30	38.5	18.7	19.8	80	19.8	7.6	12.2
31	38.3	18.4	19.9	81	18.2	6.7	11.4
32	39.9	19.3	20.6	82	13.5	5.0	8.5
33	40.0	19.3	20.8	83	11.0	3.8	7.2
34	41.0	19.7	21.3	84	10.6	3.5	7.1
35-39	**212.1**	**103.7**	**108.4**	**85-89**	**39.1**	**11.6**	**27.5**
35	41.6	20.3	21.3	85	9.7	3.1	6.6
36	42.7	20.9	21.8	86	9.1	2.8	6.3
37	43.0	21.0	22.0	87	8.0	2.3	5.7
38	42.7	20.8	21.9	88	6.7	1.9	4.7
39	42.0	20.7	21.4	89	5.6	1.5	4.1
40-44	**196.2**	**96.0**	**100.2**	**90 and over**	**19.6**	**4.3**	**15.3**
40	40.9	20.0	21.0				
41	39.4	19.3	20.1				
42	39.0	19.1	19.9	**Under 16**	**586.7**	**301.1**	**285.6**
43	38.8	18.9	19.9				
44	38.0	18.6	19.4	**Under 18**	**662.0**	**339.4**	**322.6**
45-49	**184.7**	**90.9**	**93.8**	**16-44**	**1,087.0**	**533.6**	**553.4**
45	36.7	18.0	18.7				
46	36.2	17.9	18.3	**45-64/59**[1]	**646.0**	**357.4**	**288.6**
47	37.4	18.4	19.0				
48	37.3	18.3	19.0	**65/60**[2] **and over**	**583.5**	**212.0**	**371.5**
49	37.1	18.3	18.8				

[1] 45-64 for males; 45-59 for females
[2] 65 and over for males; 60 and over for females

Table A5 Mid-2001 population estimates: estimated resident population by single year of age and sex

Northern Ireland
thousands

Age	Persons	Males	Females	Age	Persons	Males	Females
All ages	**1,689.3**	**824.4**	**864.9**				
0-4	**114.7**	**59.1**	**55.7**	**50-54**	**98.2**	**48.4**	**49.8**
0	21.5	11.1	10.4	50	19.4	9.7	9.7
1	22.3	11.5	10.8	51	19.6	9.7	9.9
2	23.1	11.8	11.3	52	19.5	9.6	9.9
3	23.6	12.2	11.4	53	19.6	9.6	10.0
4	24.3	12.4	11.8	54	20.0	9.7	10.2
5-9	**122.9**	**63.0**	**59.8**	**55-59**	**89.4**	**43.9**	**45.5**
5	24.0	12.4	11.5	55	18.4	9.1	9.3
6	24.1	12.4	11.7	56	18.2	8.9	9.3
7	24.3	12.4	12.0	57	18.4	9.0	9.3
8	24.8	12.6	12.2	58	18.1	8.8	9.3
9	25.7	13.2	12.5	59	16.3	8.0	8.4
10-14	**132.4**	**67.9**	**64.6**	**60-64**	**73.7**	**35.5**	**38.2**
10	26.1	13.4	12.7	60	15.0	7.3	7.7
11	26.0	13.3	12.7	61	14.8	7.1	7.6
12	26.3	13.5	12.8	62	14.9	7.2	7.8
13	26.9	13.7	13.1	63	14.5	7.0	7.5
14	27.1	13.9	13.2	64	14.5	6.9	7.6
15-19	**129.9**	**66.0**	**63.8**	**65-69**	**65.5**	**30.5**	**35.0**
15	27.1	13.9	13.2	65	14.0	6.6	7.4
16	27.2	13.9	13.3	66	13.3	6.3	7.0
17	26.4	13.4	13.0	67	12.9	6.0	6.9
18	25.5	13.1	12.5	68	12.8	5.8	7.0
19	23.7	11.8	11.8	69	12.5	5.8	6.8
20-24	**110.4**	**55.6**	**54.8**	**70-74**	**57.9**	**25.1**	**32.8**
20	22.9	11.7	11.2	70	12.4	5.5	6.9
21	22.7	11.7	11.1	71	12.0	5.3	6.7
22	22.0	11.0	11.0	72	11.5	5.0	6.5
23	21.4	10.6	10.8	73	11.0	4.7	6.3
24	21.4	10.6	10.7	74	11.0	4.7	6.3
25-29	**114.9**	**57.0**	**57.9**	**75-79**	**46.7**	**18.7**	**28.0**
25	21.7	10.8	10.8	75	10.5	4.3	6.2
26	21.7	10.8	11.0	76	9.8	4.1	5.8
27	22.9	11.4	11.5	77	9.3	3.7	5.5
28	24.0	11.9	12.2	78	8.7	3.4	5.4
29	24.5	12.1	12.4	79	8.3	3.1	5.2
30-34	**127.9**	**63.0**	**65.0**	**80-84**	**30.5**	**11.2**	**19.4**
30	25.3	12.5	12.8	80	7.9	3.0	4.9
31	25.2	12.4	12.8	81	7.3	2.8	4.6
32	25.6	12.5	13.0	82	6.0	2.2	3.8
33	25.9	12.7	13.1	83	4.9	1.7	3.2
34	26.1	12.8	13.2	84	4.5	1.5	3.0
35-39	**130.2**	**63.9**	**66.3**	**85-89**	**16.2**	**4.7**	**11.4**
35	26.2	12.9	13.2	85	4.1	1.3	2.8
36	26.8	13.2	13.7	86	3.8	1.1	2.6
37	26.4	12.9	13.4	87	3.3	1.0	2.3
38	25.8	12.7	13.1	88	2.8	0.8	2.0
39	25.1	12.2	12.9	89	2.3	0.6	1.7
40-44	**117.8**	**57.6**	**60.2**	**90 and over**	**7.3**	**1.6**	**5.7**
40	24.8	12.2	12.7				
41	24.2	11.8	12.3				
42	23.2	11.3	11.9	**Under 16**	**397.2**	**203.9**	**193.3**
43	23.2	11.4	11.9				
44	22.3	10.9	11.4	**Under 18**	**450.7**	**231.1**	**219.6**
45-49	**102.9**	**51.8**	**51.1**	**16-44**	**704.0**	**349.2**	**354.8**
45	21.8	10.9	10.9				
46	20.8	10.6	10.2	**45-64/59[1]**	**326.0**	**179.6**	**146.4**
47	20.5	10.4	10.1				
48	20.3	10.2	10.1	**65/60[2] and over**	**262.2**	**91.8**	**170.4**
49	19.5	9.8	9.7				

1 45-64 for males; 45-59 for females
2 65 and over for males; 60 and over for females

Table A6 Mid-2001 population estimates: estimated resident population by single year of age and sex

Scotland
thousands

Age	Persons	Males	Females	Age	Persons	Males	Females
All ages	**5,064.2**	**2,433.7**	**2,630.5**				
0-4	**276.3**	**141.8**	**134.5**	**50-54**	**350.9**	**173.9**	**176.9**
0	52.0	26.3	25.7	50	66.0	32.7	33.3
1	53.1	27.4	25.6	51	66.8	33.2	33.7
2	55.7	28.6	27.1	52	68.9	34.3	34.7
3	56.8	29.1	27.7	53	71.4	35.4	36.0
4	58.7	30.3	28.4	54	77.7	38.4	39.3
5-9	**305.8**	**156.4**	**149.5**	**55-59**	**290.1**	**142.0**	**148.2**
5	58.1	29.8	28.3	55	60.2	29.8	30.4
6	59.1	30.3	28.8	56	57.8	28.4	29.4
7	61.0	31.1	29.9	57	59.6	29.3	30.3
8	62.6	32.0	30.6	58	58.2	28.2	29.9
9	65.0	33.2	31.9	59	54.4	26.2	28.2
10-14	**322.9**	**165.7**	**157.2**	**60-64**	**261.6**	**124.6**	**136.9**
10	65.2	33.4	31.8	60	51.6	24.6	26.9
11	63.7	32.7	31.0	61	53.5	25.6	27.9
12	63.4	32.5	30.8	62	53.0	25.3	27.8
13	65.6	33.8	31.8	63	52.5	25.1	27.4
14	65.0	33.3	31.7	64	51.0	24.0	27.0
15-19	**317.6**	**161.1**	**156.5**	**65-69**	**239.5**	**110.3**	**129.2**
15	65.4	33.4	32.0	65	50.4	23.7	26.8
16	65.4	33.3	32.1	66	49.2	22.9	26.4
17	61.8	31.5	30.3	67	47.4	21.9	25.5
18	61.1	31.0	30.2	68	46.0	21.0	25.0
19	63.9	32.0	31.9	69	46.4	20.8	25.6
20-24	**315.4**	**157.7**	**157.7**	**70-74**	**207.2**	**90.2**	**116.9**
20	68.3	34.4	33.9	70	45.1	20.1	25.0
21	68.0	34.1	34.0	71	43.7	19.1	24.5
22	64.5	32.4	32.1	72	41.9	18.2	23.6
23	58.5	29.0	29.5	73	38.9	16.8	22.1
24	56.0	27.8	28.2	74	37.7	16.0	21.7
25-29	**314.9**	**153.1**	**161.8**	**75-79**	**165.6**	**66.2**	**99.4**
25	59.4	29.3	30.1	75	37.4	15.6	21.7
26	59.8	29.0	30.8	76	34.7	14.2	20.5
27	60.9	29.7	31.2	77	33.2	13.3	20.0
28	65.3	31.8	33.5	78	30.3	11.7	18.6
29	69.5	33.3	36.1	79	30.1	11.5	18.7
30-34	**381.2**	**184.2**	**197.1**	**80-84**	**106.1**	**37.0**	**69.2**
30	73.5	35.4	38.1	80	28.9	10.7	18.3
31	73.3	35.4	37.9	81	28.2	10.1	18.1
32	76.4	36.9	39.5	82	18.3	6.2	12.1
33	78.1	37.5	40.6	83	15.6	5.1	10.5
34	80.0	39.0	41.0	84	15.1	4.9	10.2
35-39	**403.2**	**194.7**	**208.5**	**85-89**	**59.4**	**16.7**	**42.7**
35	78.8	37.8	41.0	85	14.6	4.3	10.3
36	81.6	39.5	42.1	86	13.9	4.1	9.8
37	81.5	39.1	42.4	87	12.3	3.4	9.0
38	81.3	39.1	42.2	88	10.2	2.8	7.4
39	80.0	39.1	40.9	89	8.4	2.1	6.3
40-44	**378.9**	**184.6**	**194.3**	**90 and over**	**29.4**	**6.4**	**23.0**
40	78.7	38.2	40.5				
41	76.7	37.5	39.2				
42	76.1	37.0	39.1	**Under 16**	**970.4**	**497.3**	**473.1**
43	74.6	36.3	38.3				
44	72.8	35.6	37.2	**Under 18**	**1,097.6**	**562.1**	**535.5**
45-49	**338.2**	**167.2**	**171.0**	**16-44**	**2,045.9**	**1,002.0**	**1,043.9**
45	70.9	35.0	35.9				
46	68.3	33.7	34.6	**45-64/59[1]**	**1,103.9**	**607.8**	**496.1**
47	67.9	33.5	34.4				
48	66.6	33.0	33.6	**65/60[2] and over**	**944.1**	**326.7**	**617.4**
49	64.5	31.9	32.6				

1 45-64 for males; 45-59 for females
2 65 and over for males; 60 and over for females

Appendix A Notes to tables (sources, methods and definitions)

This appendix describes the data sources, calculations and definition of terms used in this volume. If further information is required, see **Appendix C** for further publications of interest and **Appendix D** for relevant contacts.

Areas covered

The statistics in this volume relate to areas with boundaries as defined on 1 April 2001.

Population estimates (Sections 1 and 2)

The figures given are the mid-year estimates of the resident population for local and health authority areas of the United Kingdom. The estimated population of an area includes all those usually resident there, whatever their nationality. Members of HM and non-UK armed forces stationed in United Kingdom are included, but those stationed outside are not. Students are taken to be resident at their term-time address. The estimates in this volume are based on the 2001 Census of Population and include allowance for a small amount of population change due to births, deaths and net migration between Census Day and mid-2001.

Population change (Section 3)

Table 3.1 and 3.2 show population change over the decades 1981-1991 and 1991-2001. Total population change has been computed using mid-1991 population estimates that have been revised in light of the results of the 2001 Census. This total population change is broken down into two components: natural change and the residual which is all other causes of population change. This other changes column provides an implied net migration estimate for most areas, but care is needed in interpreting it as such in areas with special population groups such as armed forces. In addition Migration estimates at local authority level that have been revised in light of the 2001 Census are not available. Therefore, for this volume of KPVS only, the implied net migration figures given in **Section 3** may not be consistent with net migration estimates shown in **Section 5**.

The natural change component gives the difference between numbers of live births and deaths for the period specified. Birth and death statistics are described in the notes on vital statistics **(Section 4)** below. Birth and death data presented in **Section 4** relate to the calendar year, while the natural change component of population change shown in **Section 3** relates to the period mid-1991 to mid-2001.

For the period 1981-1991 for England and Wales revised estimates of natural change and migration and other changes are not available separately.

Vital statistics, fertility and mortality rates/ ratios (Section 4)

Births

Birth statistics are compiled annually for each local government area and health area from the particulars collected at birth registration for entry into the live birth and stillbirth registers, and from the additional confidential particulars collected at the same time. This information is then entered into the live birth and stillbirth registers. For England, Scotland and Wales the confidential information is collected under the Population (Statistics) Acts of 1938 and 1960. For Northern Ireland, it is collected under Article 3(3) of the Births and Deaths Registration (Northern Ireland) Order 1976.

The numbers of births given in **Section 4** for England and Wales are those which occurred in 2001 and which were registered by 25 February 2002, plus those which occurred in 2000 but were registered in the 12 months after 25 February 2001. Some births for the year 2001 which were registered after 25 February 2002 are omitted, but these are broadly offset by the inclusion of births occurring in 2000 which were registered late. This cut off date has been extended from the original date of 11 February of the following year, which had been used since 1994. The reason for extending the date is to allow increased capture of late registrations of births. This means that annual figures are prepared using figures as close as possible to true occurences, thus providing a more accurate denominator for calculating infant mortality rates. The figures presented for Scotland and Northern Ireland are those births registered in the calendar year 2001.

For England and Wales, births are assigned to areas according to the usual residence of the mother at the date of the birth, as stated at registration. If the address of usual residence is outside England and Wales, the birth is included in any aggregate for England and Wales (and hence the UK total), but excluded from the figures for any individual region or area. In 2001 there were 274 such live births. For Scotland, births are allocated to the area of usual residence of the mother, if this is in Scotland, and to the area of occurrence if the mother's usual residence is outside Scotland. In 2001, there were 229 births in Scotland to non-resident mothers. All figures for Northern Ireland exclude births to non-resident mothers. There were 289 such births in Northern Ireland in 2001. Prior to 1998, births in Northern Ireland to non-resident mothers had been included on the same basis as in Scotland. The Northern Ireland Statistics and Research Agency (NISRA) then revised their birth figures (back to 1981) by excluding births to non-resident mothers because the majority occur in border areas and are likely to be births to residents of the Republic of Ireland. At the time, it was decided that these births should continue to be included in the total for the United Kingdom, to maintain comparability with the figures for England and Wales and for Scotland. This was the basis of the presentation of the UK figures in editions of KPVS up to and including 1999. However, these figures were not consistent with the Northern Irish birth figures used in population estimates for the UK. As a result, it was decided that from the 2000 volume onwards births registered in Northern Ireland by mothers usually resident elsewhere will be excluded from the UK figures.

A communal establishment is recorded as the mother's usual place of residence only if no other address is ascertainable or if she normally lives there - for example, if she is a member of the resident staff or a permanent hotel resident.

Deaths

The figures for deaths presented in **Section 4** for England and Wales relate to the number occurring in the calendar year 2001. A death is normally assigned to the area of usual residence of the deceased. If this is outside England and Wales, the death is included in any aggregate for England and Wales as a whole (and hence the UK total), but excluded from the figures for any individual region or area. In 2001 there were 1,286 such deaths.

For Scotland and Northern Ireland, the numbers of deaths are those registered in 2001. Deaths of Scottish and Northern Irish residents have been allocated to the area of usual residence, while deaths of non-residents have been allocated to the area of occurrence. In 2001, there were 440 deaths in Scotland of people whose usual address was outside Scotland, and 78 deaths in Northern Ireland of people whose usual address was outside Northern Ireland.

Details of the usual residence of the deceased are supplied by the informant at death registration. Where the death occurred in a communal establishment, the informant will decide whether the deceased was regarded as living there, or at some other address. However, when someone dies in a communal establishment in Scotland, and they have been there for twelve months or more, the hospital or home is automatically deemed as their normal place of residence. Further details may be found in the annual reference volume DH1 (2000), section 2.5 (on how deaths are handled in England and Wales).

Stillbirths

The definition of a stillborn child was changed on 1 October 1992 to the following:

A child which has issued forth from its mother after the twenty-fourth week of pregnancy and which did not at any time after being completely expelled from the mother breathe or show any signs of life.

Before 1 October 1992 the definition was similar except that it related to babies born dead after the twenty-eighth week of pregnancy.

Fertility rates

A basic measure of fertility is the *age-specific fertility rate*, that is, the number of live births per thousand women at each age. Age-specific rates can be calculated for single years of age, but are usually published for five-year age groups, within the reproductive age range.

The total fertility rate (TFR) provides a single measure of fertility, by summing age-specific rates for each single year of age in the reproductive age range. The TFR represents the average number of children which a woman would bear, if the current age-specific fertility persisted throughout her childbearing years.

This rate is a period measure as it refers to a period of time (usually one year) over which fertility is measured. It is a useful measure because it can be used to examine both changes in fertility over time and between populations by removing the effect of different age distributions in the female population. TFRs are shown in **Tables 4.1- 4.3** of this volume, and provide a useful means of comparing fertility between areas.

For the United Kingdom as a whole, the TFR is calculated as ΣF_j, where F_j is the age-specific fertility rate (live births per woman) at age j in the United Kingdom, and j is a single year of age in the range 15 to 44 years.

At the subnational level the TFR is calculated using the method outlined above. This is different to what has been published previously, where the TFR was calculated in five year age-groups rather than single year. For Northern Ireland TFRs are based on aggregations of three years worth of data and continue to be calculated in 5 year age-groups.

Births to women aged under 15 are included in the calculation of the age-specific rate for 15 year olds (15 to 19 year-olds for local areas), and those to women aged over 44 in the calculation of the age-specific rate for 44 year olds (40 to 44 year olds for local areas).

The percentage of births outside marriage is calculated as live births outside marriage per 100 live births. Similarly, the percentage of births outside marriage which are jointly *registered at the same address* is calculated as live births outside marriage which are registered by both parents, who give the same address of usual residence, per 100 live births outside marriage.

Births inside marriage, generally speaking, refer to those where, according to information given when the birth was registered, the parents were lawfully married to one another, either (a) at the date of the child's birth, or (b) when the child was conceived, even if they later divorced or the father died before the child's birth.

Conceptions are derived by combining registration records for both live births and stillbirths with records of legal terminations under the 1967 Abortion Act. Pregnancies leading to spontaneous abortions (i.e. miscarriages) are not included. Since many of the births in 2001 and some of the abortions were the result of conceptions in 2000, the latest data available are for 2000.

Mortality rates and ratios

To compare levels of mortality between different areas it is helpful to remove the effects of varying population structures among these areas. This can be done by using a standard set of sex and age-specific mortality rates and applying these to the population of each area. This will produce an *expected* number of deaths - that is, the number expected if the standard rates had applied in the area. This expected number is then compared with the number of deaths actually *observed* in the area.

In this volume, rates for the United Kingdom in 2001 are taken as the standard. The *standardised mortality ratio* (SMR) for each area is then calculated as (observed deaths/expected deaths) x 100

- where expected deaths = $\Sigma P_k M_k$ and
- P_k is the resident population in age/sex group k in the area
- M_k is the mortality rate (deaths per person) in age/ sex group k in the United Kingdom
- k is the age/sex group 0, 1-4, 5-14, 15-24, 25-34, 35-44, 45-54, 55-64, 65-74, 75-84, 85 and over, for males and for females.

Note that the SMRs are given for persons, with the summation (Σ) over both sexes and all ages.

Stillbirth rates are calculated as the number of stillbirths per thousand live births and stillbirths.

Perinatal mortality rates are calculated as the number of stillbirths, plus the number of deaths under one week, per thousand live births and stillbirths.

Infant mortality rates are calculated as the number of deaths under one year, per thousand live births.

Migration (Section 5)

The migration data presented in **Section 5** relate to mid-year 2000-2001.

The migration tables published in this volume are subject to revision and have been published to provide users with a **provisional** migration data series to be used until such time as the revised figures are available. They are **not** consistent with the estimated components of population change shown in **Section 3**. The revised migration figures will be made available on the ONS website (**www.statistics.gov.uk**) in late Summer 2003. Printed copies will also be available on demand and will be free of charge (email: **migstatsunit@ons.gov.uk** or telephone: 01329 813897).

The sources of the data presented in this section are the National Health Service Central Register (NHSCR) for **Tables 5.2a**, **5.2b** and **5.3**, and the NHSCR combined with patient registers held by health authorities (HAs) for **Table 5.1a** and **5.1b**. The NHSCR is notified when a patient in England and Wales transfers to a new NHS doctor in a different health authority, or health board in Scotland or Northern Ireland. Counts of these re-registrations are used as proxy indicators for movement of the population within the United Kingdom.

Every HA holds a register of the patients registered with general practitioners (GPs) within their area of responsibility. This contains the NHS number, sex, date of birth, date of acceptance by the HA and importantly the postcode of address, for each patient. By obtaining an annual download from each patient register and by combining all the patient register extracts together, a total register for the whole of England and Wales can be created. Comparing records in one year with those of the previous year by linking on NHS number enables identification of people who have changed address. The snapshot approach of the patient-register migration data means that certain groups of migrants who are missing from the register in either of two consecutive years will not be captured (see note (ii) below). Therefore, to adjust for these limitations, the data are constrained to the more complete NHSCR data. These figures are presented in **Tables 5.1a** and **5.1b**.

Table 5.2, showing migration data by areas of origin and destination, has been separated into two sections. **Table 5.2a** shows data for sub-divisions of Government Office Regions (GORs) into metropolitan counties and the remainder of the region, or the whole region if it does not contain a metropolitan county. **Table 5.2b** shows data for each region as a whole. Moves within the area are also shown, in bold type, in the diagonal, to distinguish them from the moves across the boundary of the area. Within-area moves are moves between the HAs in the region for **Table 5.2b** and between the HAs in the metropolitan county or remainder of the region for **Table 5.2a**.

Certain characteristics of the NHSCR and patient-register data should be noted:

(i) Both data sources record movements of patients as they move house and change their GP, so do not include migratory movements of people who are not registered with an NHS GP, in particular armed forces personnel and their dependants.

(ii) The patient-register data compares two snapshots to identify migrants, so certain types of moves will be missed. This is because it cannot capture the movement of migrants who were not registered with a doctor in one of the two years, but who moved during the year. The largest group of these is migrant babies aged less than one year, who would not be on a register at the start of the year. Other people, who are not on the register at the start of the year but who move after joining the NHS and before the end of the year, would not be captured. Such people could include those leaving the armed forces or international in-migrants. Similarly, people who move within the year but are not on a register at the end of the year would also not be captured. Such people would include anyone who moved and then, before the end of the year, either died or enlisted in the armed forces or left the country. However, all these within-year moves are included in the existing migration estimates derived from the NHSCR, so this more complete information is combined with the more geographically detailed data from the patient registers, to produce migration estimates for local and health authority areas.

(iii) The lowest geographical level at which data is presented in these tables is the health authority or local authority. Therefore, moves of short distances within a health or local authority are not included in the figures.

(iv) There is variation in the period between a person moving and the subsequent re-registration. Currently, the assumption is that it takes one month for a person to re-register with a GP and hence appear on a register after they move to a new area.

Table 5.3 shows moves to/from the countries of the UK and regions of England from/to elsewhere in the UK. Migration data for moves to Scotland and Northern Ireland are provided by the Northern Ireland Statistics & Research Agency (NISRA) and the General Register Office for Scotland (GROS), and these figures are agreed by the three offices with reference to the NHSCR. However, data on migration into England and Wales from Scotland and Northern Ireland are obtained directly from the NHSCR and also agreed between the three offices. In volumes up to 1998, moves to/from outside the UK were derived only from the International Passenger Survey (IPS), a continuous voluntary sample survey covering the principal air, sea and Channel Tunnel routes between the UK and overseas. The IPS excludes persons seeking asylum after entering the country and short-term visitors granted extensions to stay for a year or more. It also excludes routes between the UK and the Irish Republic. In 1999 and 2000 international migration figures represented total international migration, using data from the Home Office and the Irish Central Statistics Office (CSO) to incorporate those migrants previously excluded. This year a revised methodology for producing Total International Migration estimates is being developed. No estimates for Total International Migration have therefore been included in this volume, but will be included in revised tables on the web and in print on demand copies.

Symbols and conventions used in this volume

• not applicable	:
• not available	..
• negligible (less than half final digit shown)	-
• zero	0
• provisional	P
• revised	R

Rates calculated from fewer than 20 cases are distinguished by italic type as a warning to the user that their reliability as a measure may be affected by the small number of events.

Changes in age bands

The age-bands used in **Table 1**, **Table 2.1**, **Table 2.2** and **Table 5.1**, with age-groups 5-15 and 16-29, reflect the ages of compulsory education. These new age-bands appeared for the first time in the 1998 edition of KPVS. Previously the age-groups 5-14 and 15- 29 were used. Other age-bands in these tables are unaffected by this change.

In **Tables 4.1**, **4.2** and **4.3** the age-group for teenage conceptions refers to under 18 year-olds, as this is the age-group to which the Government's target to reduce the number of teenage conceptions relates. (Prior to 1999 edition of KPVS teenage-conceptions included conceptions to under 20 year-olds.)

Appendix B Presentation of areas

Changes that have occurred in local government and health authority organisation are briefly covered in the **Introduction**, and full details of changes occurring between 1996 and 1 April 1998 are given in the ONS printed and electronic publication: *Gazetteer of the old and new geographies of the United Kingdom* (ONS, 1999). This appendix gives information on the most recent changes, which affect the geographic areas presented in this volume of KPVS.

The data shown in the tables cover both UK constituent countries and local and health authorities, presented according to ONS and GSS guidelines. Figures are shown for all 2001 local and health authorities where available.

Further information on the presentation of statistics by area may be obtained from ONS Geography.

Presentation of local authorities: former counties, abolished counties and other areas

In recent editions, prior to this volume, figures were also presented for former counties and other groupings. These included counties which no longer exist or which had been subject to boundary change and reorganisation, particularly where unitary authorities were created from parts of the former county.

Only one non-standard area is now presented. This can be cross-referenced with **Table B.1** shown below detailing the current area which the 'non-standard' area covers.

No changes were made to the structure of local authorities in the year up to 1 April 2000, and therefore the presentation of current local authorities is the same as those shown in the 2000 edition of Key Population and Vital Statistics.

Health authority areas

This section describes the most recent changes to health areas, which were made on 1 April 2001.

Health regional office areas

No changes were made to health regional office areas in the year up to 1 April 2001

Health authority areas

There were eight changes made to the structure of health authorities (HAs) on 1 April 2001. The areas affected are detailed in **Table B.2**.

Table B.1 Notes to 'non-standard' area presented in this volume, and the area covered.

Area Referred to in Volume	Equivalent area in current local government structure
The Isles of Scilly	The Isles of Scilly are separately administered by an Isles of Scilly Council and do not form part of the county of Cornwall, but are usually associated with the county.

Table B.2 Changes to regional health areas from 1 April 2001

Regional health office areas	Health authorities from 1 April 2001	Former Health authorities
South East	Isle of Wight, Portsmouth and South East Hampshire	Portsmouth and South East Hampshire Isle of Wight
London	Barnet, Enfield and Haringey	Barnet & Enfield Haringey
Eastern	Hertfordshire	East and North Hertfordshire West Hertfordshire
London	Bexley, Bromley and Greenwich	Bexley and Greenwich Bromley

Appendix C Other relevant publications

Data availability on the internet

The internet is becoming an increasingly important means of disseminating National Statistics data. A large proportion of publications are now available free on the National Statistics website, and via TSO's print-on-demand service (a charge is made for this service).

Statbase is the Government Statistical Service (GSS) database set up to provide customers with a comprehensive set of key statistics. Statbase may be accessed via the internet by clicking on the National Statistics website (**www.statistics.gov.uk**). For tables in the quarterly publications, or volumes, click on either 'Publications' or follow the topic links.

Population and vital statistics for earlier issues

Statistics of population, births and deaths, and since 1985 of migration, have appeared in previous issues of this annual publication. For the years 1974 to 1982 KPVS was published as *Local Authority Statistics* (Series VS nos 1 to 9), for 1983 as *Vital statistics: local and health areas* (Series VS no 10), for 1984 and 1985 as *Population and vital statistics: local and health authority area summary* (Series VS no 11/PP1 no 7 and VS no 12/PP1 no 8) and from1986 as *Key population and vital statistics: local and health authority areas* (Series VS nos 13-26/PP1 nos 9-22). In the 1997 volume, data for Scotland and Northern Ireland were also included.

Following the 2001 Census of Population, population estimates have been revised for each year 1982 to 2000 for England, Wales, Scotland and for 1992 to 2000 for Northern Ireland. These population estimates are available on the National Statistics website.

Readers wishing to obtain detailed local or national population estimates for earlier years or information about the formats in which they are available should contact the Population Estimates Unit, ONS, at the address given on page 118.

Statistics for **births and deaths** in England and Wales for 1973 and earlier years were published in the *Registrar General's Statistical Review of England and Wales*.

Similar statistics for Scotland and Northern Ireland are published in *Scotland's Population 2001 – The Registrar General's Annual Review of Demographic Trends* and the *Annual Report of the Registrar General for Northern Ireland* respectively.

Unpublished vital statistics tables

The content of the vital statistics tables for England and Wales available on CD ROM in previous years is currently under review for new data years. These tables provided detailed statistics on births and deaths for local authorities and in some cases wards. Their content is being reviewed to ensure they comply with the principles on protecting confidentiality in the National Statistics Code of Practice. In the meantime, tables for new data years giving fewer details, but for the same geographical areas will be available from Vital Statistics Outputs, ONS Titchfield. Information about the tables can be obtained from the address in Appendix D.

Reference publications

Statistics for local areas appear in many National Statistics publications and on the National Statistics website (**www.statistics.gov.uk**). The following annual volumes have been published recently or are about to be published.

Migration
International Migration, 2000 (Series MN no. 27, The Stationery Office, 2002)

Time series and annual data from the International Passenger Survey, Home Office data and data from the Central Statistical Office (CSO) in Ireland. Includes analysis by region of destination or origin. The tables published in volume MN no.27 are subject to revision in light of the Census results, as are the tables published in this volume. MN no. 28 will include revised estimates and will be published late Summer 2003.

Family statistics
Birth statistics 2001 (Series FM1 no. 30, available on the National Statistics website: **www.statistics.gov.uk**)
Includes tables for government office regions, metropolitan counties, regional office and health authority areas showing live and still births in 2001 by occurrence inside/outside marriage, age of mother and mother's country of birth, age-specific and total fertility rates, and maternities by place of confinement. Also shown for 2000 are conception rates and the proportion and outcome of conceptions occurring inside/outside marriage.

Marriage, divorce and adoption statistics, 2000 (Series FM2 no. 28.
Marriages by area of occurrence (regions and selected administrative areas), by age, sex and previous marital status.

Deaths

Mortality statistics: general 2000 (Series DH1 no. 33, 2002).
Deaths by sex and age, by certain characteristics collected at the time of registration, and by place of death, usual residence and country of birth of deceased. It also contains tables on expectation of life, and on years of life lost due to certain causes of death. Several of the tables also show time trends.

Mortality statistics: cause 2001 (Series DH2 no. 28).
Deaths by underlying cause, including age-specific rates, age-standardised mortality rates and time trends in mortality. It also includes tables analysing all conditions mentioned on the death certificate.

Mortality statistics: childhood, infant and perinatal, 2001 (Series DH3 no. 34).
Data on stillbirths, infant deaths and childhood deaths. It also includes figures on infant deaths linked to their corresponding birth records.

Mortality statistics: injury and poisoning, 2001 (Series DH4 no. 26).
Accidental and violent deaths by place of accident, nature of injury, cause of accident and verdict at coroner's inquest.

Morbidity

Cancer statistics: registrations, 1999 (Series MB1 no. 30).
Newly diagnosed cases of cancer for urban/rural areas and RHAs by age, sex and site.

Annual review of Communicable disease statistics, England and Wales, 1999/2000 (Public Health Laboratory Service, 2001).
Statutory notifications of infectious diseases for standard regions and regional health offices by age-group and sex and totals for local government areas. Laboratory identifications of communicable disease organisms by age-group of regional health office. Editions for previous years published by HMSO as *Communicable disease statistics* (Series MB2 nos 1 to 21).

Congenital anomaly statistics, 2000 (Series MB3 no. 16, The Stationery Office, 2002).
Notifications of specific malformations by standard region, regional/district health authority, sex, age of mother and social class.

Abortion statistics, 2001 (Series AB no. 28, The Stationery Office, 2002).
Legally induced abortions for health areas by age of mother, marital status, previous children, gestation period, statutory grounds, category of premises, operation and duration of stay.

Scotland

Scotland's Population 2001 – The Registrar General's Annual Review of Demographic Trends (General Register Office for Scotland, 2002)
The Annual Review of the Registrar General for Scotland presents statistical information derived from the registration of 'vital events' - births, deaths and marriages, together with divorces and adoptions - and related statistical information on the population of Scotland. In addition to the year in question, time-series of data are also included.

Scotland's Census 2001 – 2001 Population Report (General Register Office for Scotland, 2002).
Population estimates by sex, quinary age group and administrative areas (council and health board areas) are included.

Population Projections, Scotland (2000-based) (General Register Office for Scotland, 2002).
Projected populations by sex, age and administrative area for the period up to 2016 for council and health board areas.

Northern Ireland

Annual Report of the Registrar General, 2001 (The Stationery Office, 2002).
This report contains vital statistics on the births, deaths and marriages registered in Northern Ireland during the year ended 31 December 2001 and on adoptions and divorces which took place during the same period.

The report contains nine chapters on population; births; stillbirths and infant deaths; causes of death; marriages; divorces; adoptions and re-registrations. Each chapter contains key points, commentary, graphs and tables. A summary chapter at the beginning of the report draws all this information together. Notes and definitions are contained in the Appendices of the report along with a report on the work of the General Register Office for Northern Ireland. A special report on demographic profile of parliamentary constituencies is also included. The report is available, at a cost of £25.00, from The Stationery Office bookshop.

Northern Ireland Annual Abstract of Statistics, 2002 (The Stationery Office, 2002).
The Abstract is compiled and produced by the Northern Ireland Statistics and Research Agency and contains a wide range of statistics on inter alia, demography, health, the environment, the economy, agriculture, security, justice and education. The Abstract may be obtained at a cost of £20.00 from the Stationery Office bookshop.

A detailed account of the Northern Ireland population estimation methodology is contained in the NISRA

Occasional Paper No 12 entitled - *Mid-year population estimates in Northern Ireland - Validation and extension to local government districts* - available, at a cost of £10, from:

Corporate Branch
Northern Ireland Statistics and Research Agency
McAuley House
2 - 14 Castle Street
Belfast
BT1 1SA
Telephone: 02890 348114
Fax: 02890 348117

2001 Census of Population

England and Wales

The results of the 2001 Census are available in three ways:

1. printed reports available through The Stationery Office;
2. electronic media available on request, from Census Customer Services (contact details - see **Appendix D);**
3. ONS website - **www.statistics.gov.uk/census2001**.

The first published report - *First results on population for England and Wales* and data from the 2001 Census were released on 30 September 2002. The report includes statistics for the population on Census Day 2001 by single year of age and sex and the population by quinary age groups for local authority districts, unitary authorities, counties and Government Office Regions.

Key Statistics for Local Authorities in England and Wales was released on 13 February 2003 in both published and electronic formats. It provides the first opportunity to see a summary of the complete results of the Census, set out to enable comparison at local, regional and national level. A bilingual report for Wales was released at the same time. Further information regarding developments and schedule of release for future datasets can be viewed within the Output Prospectus on the National Statistics website (**www.statistics.gov.uk/Census2001/op.asp**).

Northern Ireland

Standard tables for the 2001 Census for Northern Ireland are available. These are available on the NISRA website: **www.nisra.gov.uk/Census**

A list of scheduled release dates for further information can be found on the web at : **www.nisra.gov.uk/census/pdf/outputprospectus.pdf**

Further information on the Census may be available from:

Census Customer Services
Northern Ireland Statistics and Research Agency
McAuley House
2-14 Castle Street
Belfast
BT1 1SA
Tel: (028) 9034 8160
Fax: (028) 9034 8161
Email: **census.nisra@dfpni.gov.uk**

Population Trends and Health Statistics Quarterly

Up to the end of 1998 ONS published annual and quarterly data in the quarterly journal, *Population Trends*. Each issue contained a regular series of tables on population and vital statistics for standard regions and health regions, and articles on a variety of population and medical topics.

Since 1999 ONS has published two quarterly journals: *Health Statistics Quarterly*, which covers mortality and other health topics, and *Population Trends*, with an emphasis on population and demography, but including some mortality data. The first issue of *Health Statistics Quarterly* appeared in February 1999, while the first issue of *Population Trends* with a revised coverage was published in March 1999.

Reports on Population Trends and Health Statistics Quarterly

Population Trends and *Health Statistics Quarterly* contain reports which are used to supplement the ONS annual publication series and provide a rapid release of some information as it becomes available. Up until 1998 this role was carried out by a series of Monitors. With two exceptions - population estimates and electoral registers, which are detailed below - these were subsumed into one or other of these publications. Topics covered in reports in *Population Trends* and *Health Statistics Quarterly* over the past year include:

Legal abortions

Terminations carried out in England and Wales by (i) regional health office area of operation, type of premises, and planned day care (quarterly), and (ii) health authority of residence, marital status, age-group, parity, and regional health areas of operation (annual). 'Legal abortions in England and Wales, 2000', *Health Statistics Quarterly*, 12 (2001).

Deaths

Includes numbers and SMRs for deaths occurring in government office regions and local authorities, and health regional office areas and health authorities (annual). It also includes deaths by cause for England and Wales. 'Death registrations in England and Wales, 2001: area of residence', *Population Trends*, 108 (2002).

Infant and perinatal mortality

Includes infant and perinatal mortality statistics for HAs (annual). 'Infant and perinatal mortality 2001: health areas, England and Wales', *Health Statistics Quarterly*, 15 (2002).

Births and conceptions

Births by mother's residence for government office regions and local authorities, and health regional office areas and health authorities (annual).
'Conceptions in England and Wales, 2000', *Health Statistics Quarterly*, 13 (2002). 'Live births in England and Wales, 2001: local and health authority areas', *Population Trends*, 108 (2002).

Infectious diseases

Corrected summary of notified cases of infectious disease in health regional office and local government areas, and laboratory identifications of communicable diseases in health regional office areas (quarterly).

Population projections - sub-national

Population projections for regions, local government areas and health areas of England. '2000-based short term subnational population projections for health authority areas in England; *Population Trends* 108 (2002).

ONS first releases

The following data, formerly released via ONS monitors, are now issued by first releases, although reports may be published in *Population Trends* or *Health Statistics Quarterly* at a later date:

Electoral

Local government electors in the United Kingdom will be published in a first release:
UK Electoral Statistics for 2002.

Population estimates

Mid-year estimates of the resident population of the UK are published in a first release:
Mid-2001 UK Population Estimates.

Population and Health Monitors were published by ONS up to 1998. Copies can be obtained by contacting ONS Direct, Government Offices, Cardiff Road, Newport, Gwent NP9 1XG or telephone 01633 816262, or on the National Statistics website (see **Appendix D**).

Migration

On 28 November 2002 ONS issued a first release of interim revised international migration estimates for the UK for 1992-2001. These data took preliminary account of the results of the 2001 Census which shows that existing sources had overestimated the net inflow of migrants to the UK. Migration estimates for this period, including the figures in this volume, are therefore subject to revision.

Articles in population Trends and Health Statistics Quarterly

Recent articles in *Population Trends* include:

Divorce and remarriage in England and Wales, *Population Trends,* 113 (2003).

New estimates and projections of the population cohabiting in England and Wales, *Population Trends,* 95 (1999).

The demography of centenarians in England and Wales, *Population Trends,* 96 (1999).

The ethnic minority populations of Great Britain - latest estimates, *Population Trends,* 96 (1999).

Selection of topics and questions for the 2001 Census, *Population Trends,* 97 (1999).

Which authorities are alike? *Population Trends,* 98 (1999).

Families, Groups and clusters of local and health authorities: revised for authorities in 1999, *Population Trends,* 99 (2000).

2000-based national projections for the United Kingdom and constituent countries, *Population Trends,* 107 (2002).

Reviewing the mid-year population estimates in Northern Ireland, *Population Trends,* 99 (2000).

Variant population projections for the United Kingdom and its constituent countries, *Population Trends,* 109 (2002).

25 years of Population Trends, *Population Trends,* 100 (2000).

Trends in fertility and contraception in the last quarter of the 20th century, *Population Trends,* 100 (2000).

Developments in Census taking in the last 25 years *Population Trends,* 100 (2000).

Fragmented life courses: the changing profile of Britain's ethnic populations, *Population Trends,* 101 (2000).

The use of patient registers to estimate migration, *Population Trends,* 101 (2000).

Geographic variations in conceptions to women under 18 in Great Britain during the 1990s, *Population Trends,* 102 (2000).

United Kingdom population trends in the 21st century, *Population Trends,* 103 (2001).

Assumptions for the 2000-based National Population Projections, *Population Trends,* 105 (2001).

Population Review of 2000: England and Wales, *Population Trends,* 106 (2001).

International migration to and from the UK, 1975-1999, consistency, change and implications for the labour market, *Population Trends,* 106 (2001).

Implications of changes in the UK social and occupational classifications in 2001 for vital statistics, *Population Trends,* 107 (2002).

The Millenium Cohort Study, *Population Trends,* 107 (2002).

New estimates of trends in births by birth order in England and Wales, *Population Trends,* 108 (2002).

Attitudes towards ideal family size of different ethnic/nationality groups in Great Britain, France and Germany, *Population Trends,* 108 (2002).

Rebasing the annual mid-year population estimates for England and Wales, *Population Trends,* 109 (2002).

The effect of changes in timing of childbearing on measuring fertility in England and Wales, *Population Trends,* 109 (2002).

Recent articles in *Health Statistics Quarterly* include:

Trends in cot deaths, *Health Statistics Quarterly,* 05 (2000).

Trends in regional deprivation and mortality using the Longitudinal Study, *Health Statistics Quarterly,* 05 (2000).

Recording births and deaths in the countries of the United Kingdom, *Health Statistics Quarterly,* 06 (2000).

Geographic inequalities in mortality in the United Kingdom during the 1990s, *Health Statistics Quarterly,* 07 (2000).

Healthy life expectancy in Great Britain, 1980-96, and its use as an indicator in United Kingdom Government strategies, *Health Statistics Quarterly,* 07 (2000).

Analysis of risk factors for neonatal mortality in England and Wales, 1993-97: based on Singleton babies weighing 2,500-5,499 grams, *Health Statistics Quarterly,* 08 (2000).

Daily and seasonal variation in the births, stillbirths and infant mortality in England and Wales, 1979-96, *Health Statistics Quarterly,* 09 (2001).

Geographic Inequalities in Life Expectancy in the United Kingdom, *Health Statistics Quarterly,* 09 (2001).

Are unascertained deaths the same as sudden infant deaths? *Health Statistics Quarterly,* 10 (2001).

The calculation of abortion rates for England and Wales, *Health Statistics Quarterly,* 10 (2001).

Geographic Variations in deaths related to drug misuse in England and Wales, *Health Statistics Quarterly,* 11 (2001).

Linking HES maternity records with ONS birth records, *Health Statistics Quarterly,* 13 (2002).

The implementation of ICD-10 for cause of death coding – some preliminary results from the bridge coding study, *Health Statistics Quarterly,* 13 (2002).

Results of the ICD-10 bridge coding study, England and Wales, *Health Statistics Quarterly,* 14 (2002).

Causes of neo-natal deaths and still births: a new hierarchical classification in ICD-10, *Health Statistics Quarterly,* 15 (2002).

Inequalities in life expectancy by social class, 1972-1999, *Health Statistics Quarterly,* 15 (2002).

Census News

Combining the previous CEN Series of *ONS Monitors* for England and Wales and the *Census information bulletins* for Scotland, Census News aims to maintain contact between the Census Offices and census users by providing information on the 2001 Census and details of relevant services, product proposals, meetings, and other user consultation activities.

Census News is available, free of charge, from Census Customer Services, ONS, Segensworth Road, Titchfield, Fareham, Hampshire PO15 5RR or telephone 01329 813800. It can be viewed on the ONS website: **www.statistics.gov.uk**, and can be delivered in electronic form, e-mail: **census.customerservices@ons.gov.uk**

National Statistics Horizons

This is the magazine for National Statistics customers and is produced by ONS in conjunction with other government departments whose data come under the National Statistics banner. It contains a summary of the latest news from National Statistics to help and inform people in local and health authorities about production timetables, availability of data, coding procedures, and so on. The magazine is produced quarterly and is free. Contact National Statistics Customer Enquiries for further details, tel: 0845 601 3034.

Studies on Medical and Population Subjects

This series of occasional publications provides a more extensive account of important subjects than is practical within the limits of the annual publication series. Recent studies, obtainable from The Stationery Office, include:

The ONS classification of local and health authorities of Great Britain: revised for authorities in 1999, SMPS no. 63 (1999).

Death certification and mortality statistics: an international perspective, SMPS no. 64 (2000).

Medical research at the Office for National Statistics: A review, SMPS no. 65 (2000).

Cancer Trends in England and Wales 1950-1999, SMPS no. 66 (2001).

ONS Occasional Papers

This series publishes information which is likely to be of interest to a specialised readership. The range of topics and contents is varied. Although these volumes are not currently in print, they are stocked in many libraries.

Population estimates methodology, *Occasional Paper* 37 (1991) . (The updated version is available on National Statistics website: **www.statistics .gov.uk**).

British Society for Population Studies Conference Papers 1990, Population projections: Trends, methods and uses. *Occasional Paper* 38 (1990).

A review of migration data sources, *Occasional Paper* 39 (1991).

Accuracy of the rolled forward population estimates in England and Wales, 1981-1991, *Occasional Paper* 44 (1995).

Other publications

Gazetteer of the old and new geographies of the United Kingdom (ONS, 1999).

Migration within England and Wales using the ONS Longitudinal Study, LS no.9 (The Stationery Office, 2000).

Geographic Variations in Health, DS no.16 (The Stationery Office, 2001)

United Kingdom Health Statistics, UKHS no.1 (The Stationery Office, 2001).

Other publications available from The Stationery Office include *Regional Trends, Social Trends* and *Labour Market Trends.*

Appendix D Further information and point of contact

Data	Unit (and address)	Telephone/Email Address/Fax
National Statistics Website		www.statistics.gov.uk
General enquiries		
Central enquiries	Customer Enquiry Centre	0845 601 3034

The ONS Customer Enquiry Centre will forward calls to the appropriate area, including queries about publications. Customers may call this number in the first instance.

Topic enquiries

A full postal address for each area is listed at the end of the appendix

Geography Information on the presentation of statistics by area	ONS Geography **Address A**	tel: 01329 813477 e-mail: geography@ons.gov.uk fax: 01329 813383
International migration For the UK and all of the UK countries	Migration Statistics **Address A**	tel: 01329 813255 e-mail: migstatsunit@ons.gov.uk fax: 01329 813295

England and Wales

Population estimates	Population Estimates **Address A**	tel: 01329 81 3318/3453 e-mail: pop.info@ons.gov.uk fax: 01329 813295
Vital statistics	Vital Statistics Outputs **Address A**	tel: 01329 813758 e-mail: vsob@ons.gov.uk fax: 01329 813548
Subnational population projections for England	Subnational Projections **Address A**	tel: 01329 81 3865/3724 e-mail: subnatproj@ons.gov.uk fax: 01329 813295
Internal migration (population movements within England and Wales)	Migration Statistics **Address A**	tel: 01329 813872 e-mail: migstatsunit@ons.gov.uk fax: 01329 813295
Census: Orders for data from the 2001 and previous Censuses. Information on release of data from the 2001 Census.	Census Customer Services **Address A**	tel: 01329 813800 e-mail: censuscustomerservices@ons.gov.uk fax 01329 813587

Data	Unit	Telephone/Email Address/Fax (see footnote for full details)
Scotland		
Contact points for Scotland data		
Population estimates Subnational population projections Internal migration 2001 Census General Enquires	GROS Demography and Dissemination Branch **Address B**	tel: 0131 314 4254 e-mail: customer@gro-scotland.gsi.gov.uk fax: 0131 314 4696
Vital Statistics Publication Sales General Enquiries	GROS Demography and Dissemination Branch **Address B**	tel: 0131 314 4243 e-mail: customer@gro-scotland.gsi.gov.uk fax: 0131 314 4696
Conceptions and birthweight information	Customer Support Desk Information and Statistics Division SHS **Address C**	tel: 0131 551 8899 e-mail: csd@isd.csa.scot.nhs.uk fax: 0131 551 1392
Northern Ireland		
Contacts for Northern Ireland data		
Population estimates Subnational population projections Internal migration	NISRA (Demography and Methodology branch) **Address D**	02890 348132 e-mail: dmb@dfpni.gsi.gov.uk fax: 02890 348134
2001 Census of Population	Census Customer Services, NISRA **Address D**	028 9034 8160 e-mail: census.nisra@dfpni.gov.uk fax: 028 9034 8161
Births and Deaths	NISRA (GRO) **Address E**	02890 252000 fax: 02890 252044
National population projections for the United Kingdom and constituent countries	Government Actuary's Department **Address F**	tel: 020 7211 2622 e-mail: enquiries@gad.gov.uk fax: 020 7211 2640

Contact postal addresses for further information

Address A
Office for National Statistics (ONS)
Segensworth Road
Titchfield
Fareham
Hampshire
PO15 5RR

Address B
Customer Services
Demography and Dissemination Branch
General Register Office for Scotland (GROS)
Ladywell House, Ladywell Road
Edinburgh
EH12 7TF

Address C
Scottish Health Service (SHS)
Trinity Park House
South Trinity Road
Edinburgh
Scotland
EH5 4SQ

Address D
Northern Ireland Statistics
and Research Agency
McAuley House
2 - 14 Castle Street
Belfast
BT1 1SA

Address E
Northern Ireland Statistics
and Research Agency (GRO)
Oxford House
49-55 Chichester St
Belfast
BT1 4HL

Address F
Government Actuary's Department (GAD)
New King's Beam House
22 Upper Ground
London
SE1 9RJ